T0140676

Sustainable Production, Life Cycle Engineering and Management

Series editors

Christoph Herrmann, Braunschweig, Germany
Sami Kara, Sydney, Australia

Modern production enables a high standard of living worldwide through products and services. Global responsibility requires a comprehensive integration of sustainable development fostered by new paradigms, innovative technologies, methods and tools as well as business models. Minimizing material and energy usage, adapting material and energy flows to better fit natural process capacities, and changing consumption behaviour are important aspects of future production. A life cycle perspective and an integrated economic, ecological and social evaluation are essential requirements in management and engineering. This series will focus on the issues and latest developments towards sustainability in production based on life cycle thinking.

More information about this series at http://www.springer.com/series/10615

Christoph Herrmann · Mark Stephan Mennenga
Stefan Böhme
Editors

Fleets Go Green

Springer

Editors
Christoph Herrmann
Institute of Machine Tools and Production Technology (IWF)
Technische Universität Braunschweig
Braunschweig, Lower Saxony
Germany

Stefan Böhme
Institute of Machine Tools and Production Technology (IWF)
Technische Universität Braunschweig
Braunschweig, Lower Saxony
Germany

Mark Stephan Mennenga
Institute of Machine Tools and Production Technology (IWF)
Technische Universität Braunschweig
Braunschweig, Lower Saxony
Germany

ISSN 2194-0541 ISSN 2194-055X (electronic)
Sustainable Production, Life Cycle Engineering and Management
ISBN 978-3-030-10256-2 ISBN 978-3-319-72724-0 (eBook)
https://doi.org/10.1007/978-3-319-72724-0

Printed on acid-free paper

This Springer imprint is published by the registered company Springer International Publishing AG part of Springer Nature
The registered company address is: Gewerbestrasse 11, 6330 Cham, Switzerland

Series Editors' Foreword

Electrical vehicles have been viewed as an innovative technology disrupting current business models in the automotive industry with a long-term prospect of reducing greenhouse gas emissions from the transportation sector. However, they also pose challenges for public infrastructure and traffic management. Commercial vehicle fleets, in particular, are an important use case. Vehicle fleets offer good conditions for exploiting the possible advantages of electric mobility due to their use profiles and the high number of the commercial admission of new cars.

However, many questions regarding electric mobility remain unanswered today due to the lack of long-term experiences and extensive field tests. General statements about the environmental advantages of electric vehicles cannot easily be made due to the high number of influencing factors and various interdependencies. Diverse parameters need to be considered before deciding new electric vehicle concepts and being matched to the specific use. If electric vehicles are meant to exploit their whole potential for reducing emissions and minimizing the consumption of resources, it is necessary to take an integrated view of all the interdependent factors and their interactions.

The environmental, economic, and social advantages and disadvantages of electric vehicles might only be completely understood when they are analyzed from a life cycle perspective—considering the extraction of raw materials, the production of the vehicles and components as well as the actual vehicle use and the final dismantling and disposal. If powered with low carbon-intensive energy sources, electric vehicle fleets might reduce their impact on climate change. Nevertheless, the production of electric vehicles is highly material- and energy-intensive. Therefore, an electric vehicle and its components require a large amount of materials that are associated with a wide range of potential environmental impacts beyond climate change. In order to be able to support decisions for policy development, and product and business strategies on the promotion and implementation of electric vehicles into corporate fleets, it is essential to consider a life cycle perspective. Life cycle thinking is necessary to avoid conflicts of interests or problem shifting when deciding about future mobility.

The joint research project "Fleets Go Green" attempts to address these challenges. The project aims at studying the environmental impacts of electric vehicles in fleet operations from a life cycle perspective. The contributions to this book investigate the integrated behavior of the vehicle, users, and the power supply system. After a short introduction presenting the motivation and outline of the project, Chap. 2 determines the total energy requirements of fleet vehicle operations with different topologies over the use phase. The user acceptance of fleet owners and drivers is explained in Chap. 3. Chapter 4 aims at integrating electric vehicle fleets into the electrical distribution system and introduces ways to maximize the use of renewable energy sources for charging them. The environmental assessment of fleets is studied in detail in Chap. 5. Finally, all findings are integrated into a decision support system for the ecologically oriented fleet management and planning in Chap. 6. In the concluding Chap. 7, recommendations for further developing and promoting electric mobility are presented. As a result of four years of research, the authors provide the reader with valuable insights into the various interdependent factors of electric mobility and their interactions, with a special focus on its application in commercial fleet operations. Therefore, this book is a significant contribution to our understanding of electric mobility as a system of systems.

Christoph Herrmann
Technische Universität Braunschweig

Sami Kara
The University of New South Wales

Contents

Abbreviations

φ_{Act}	Actual temperature
$\varphi_{Environment}$	Temperature of the environment
AC	Alternative Current
AP	Acidification Potential
BEV	Battery Electric Vehicles
BMUB	Federal Ministry of Environment, Nature Conservation, Building and Nuclear Safety
BOM	Bill of Materials
CFC	Carbon Fiber Composite
CHAdeMO	Charge de Move
CML	Centrum voor Milieukunde der https://de.wikipedia.org/wiki/Universit%C3%A4t_Leiden "\o"Universität Leiden
CNG	Compressed Natural Gas
DC	Direct Current
ED	Electric Drive
EEA	European Environment Agency
$E_{EV,k}$	State of charge of the k-ten vehicle from the fleet
EM	Electric Motor
EP	Eutrophication Potential
Eq./eq.	Equivalent
E_{sp}	State of charge of the stationary storage
EV	Electric Vehicle
EVB	Electric Vehicle Battery
FGG	Fleets Go Green
F_{Trac}	Force of Traction
GHG	Greenhouse Gas
GWP	Global Warming Potential
HV	High Voltage
HVAC	Block for heating, ventilation, and air-conditioning
I_{Bat}	Current of the battery

ICE	Internal Combustion Engine
IEA	International Energy Agency
IFAM	Fraunhofer Institute for Manufacturing Technology and Advanced Materials
ILCD	International Life Cycle Data
IMDS	International Material Data System
KBA	Federal Office for Motor Traffic
LCA	Life Cycle Assessment
LCC	Life Cycle Costing
LCIA	Life Cycle Impact Assessment
LP	Linear Problem
LV	Low Voltage
M/G	Motor/Generator
M_{EM}	Torque of Electric Motor
M_{Trans}	Transmission-Torque
NEDC	New European Driving Cycle
n_{EM}	Speed of Electric Motor
NFF	Automotive Research Center Niedersachsen
n_{Trans}	Transmission-Speed
n_{Wheel}	Speed of the wheel
ODP	Ozone Depletion Potential
OEM	Original Equipment Manufacturer
P_{conv}	Load
PDM	Product Data Management
$P_{gen1,2,3}$	Power of energy source 1, 2, or 3
PHEV	Plug-in Hybrid Electric Vehicles
PLM	Product Lifecycle Management
P_{losses}	Losses of Power
POCP	Photochemical Ozone Creation Potential
PTC	Positive Temperature Coefficient
PV	Photovoltaik
RER	Data for Europe
SME	Small- and Medium-sized Enterprises
SoC	State of Charge
TCO	Total Cost of Ownership
T_{Trans}	Transmission-Temperature
U_{Bat}	Voltage of the battery
$v_{current}$	Current speed
VDA	Association of the Automotive Industry
VDI	Association of German Engineers
VMF	Association of brand-independent fleet management corporations
v_{target}	Target speed
$X_{EV,k}$	Charging capacity of k-ten vehicle from the fleet
$X_{in1,2,3}$	Part of electric feed in the public grid of the energy sources 1, 2 order 3
$X_{oc1,2,3}$	Proportion of consumption from energy source 1, 2 order 3

X_{sc}	Charging capacity of the stationary storage
X_{sd}	Unloading capacity of the stationary storage
X_{sup}	Electric power demand from the public grid
Z.E.	Zero Emission

Symbols

E	Energy (Wh)
F	Force (N)
I	Current (A)
M	Torque (Nm)
n	Speed (1/min)
P	Power (W)
P	*p*-value (-)
T, φ	Temperature (°C)
U	Voltage (V)
v	Speed (km/h)
α, β	Angles (°)
Δt	Time interval
η	Efficiency (%)

List of Figures

List of Tables

Chapter 1
Research for Sustainable Mobility—Fleets Go Green

Christoph Herrmann, Mark Stephan Mennenga and Stefan Böhme

Mobility is fundamental for trade and business, for science, culture and everyday life of people. An efficient transport system enables economic growth, promotes social exchange, creates more freedom and independence for each individual and thus makes a significant contribution to the quality of life. The planning and design of future mobility is associated with a multitude of fundamental challenges. Megatrends such as individualization tendencies, urbanization and aging of societies have a strong influence on future mobility concepts. In addition, technological developments like eco-efficient lightweight structures, electrification and digitalization and, last but not least, new business models on product-service systems and sharing economy lead to new vehicle and mobility concepts.

From the environmental point of view, the negative impact of mobility on human health and the ecosystem is growing with the increasing traffic volume of recent years. In Germany, around 15% of the greenhouse gas emissions result from the transport sector (see Fig. 1.1). While greenhouse gas emissions in the European Union have declined overall between 1990 and 2005, they have increased by 27% in the transport sector at the same time.[1] In order to stay within a safe operating space and to avoid peak warming above 2 °C, the industry associated with mobility will have to deliver a significant reduction of emissions.

The electrification of vehicles and the transition of the energy system towards renewable energy (German: Energiewende) is a promising strategy here. While the registration of electric vehicles has strongly increased compared to former years, in absolute terms electric and hybrid cars still represent a minority compared to the

[1]Data origin: European Environment Agency: EEA Greenhous Gas Data Viewer. http://www.eea.europa.eu/data-and-maps/data/data-viewers/greenhouse-gases-viewer.

C. Herrmann (✉) · M. S. Mennenga · S. Böhme
Institute of Machine Tools and Production Technology (IWF), Technische Universität Braunschweig, Braunschweig, Germany
e-mail: c.herrmann@tu-braunschweig.de

C. Herrmann et al. (eds.), *Fleets Go Green*, Sustainable Production, Life Cycle Engineering and Management,
https://doi.org/10.1007/978-3-319-72724-0_1

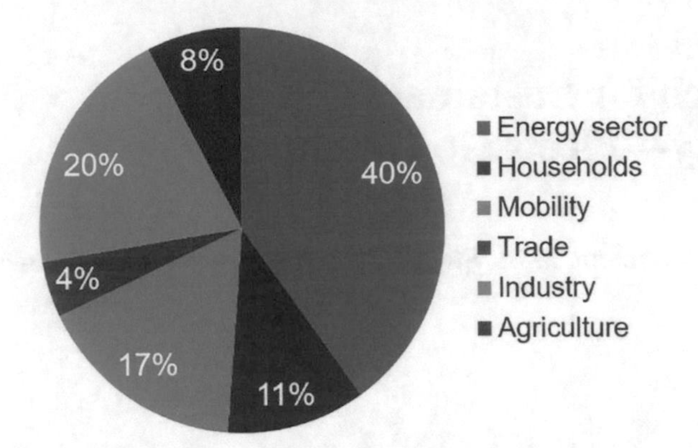

Fig. 1.1 Greenhouse gas emissions in Germany in 2015 (Data origin: Umweltbundesamt: Nationaler Inventarbericht zum Deutschen Treibhausgasinventar 1990–2015. Berichterstattung unter der Klimarahmenkonvention der Vereinten Nationen und dem Kyoto-Protokoll 2017. http://www.umweltbundesamt.de/publikationen/berichterstattung-unter-der-klimarahmenkonvention-2.)

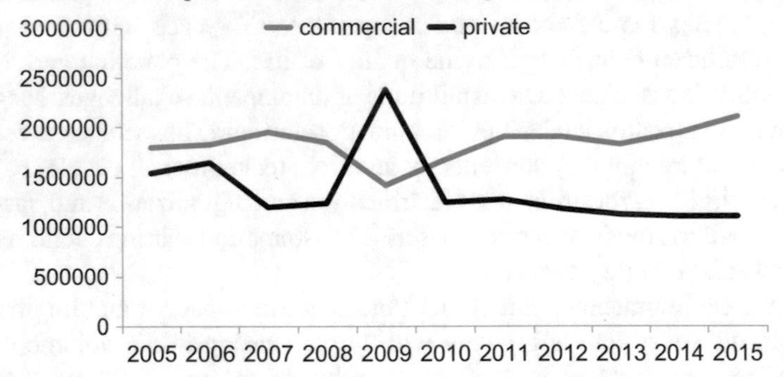

Fig. 1.2 New vehicle registrations in Germany (Data origin: Kraftfahrt-Bundesamt, see https://www.kba.de/DE/Statistik/Fahrzeuge/Neuzulassungen/Halter/z_n_halter.html?nn=652344, as well as https://www.kba.de/DE/Statistik/Fahrzeuge/Neuzulassungen/Halter/2015_n_halter_dusl.html?nn=652344.)

sum of newly registered vehicles in Germany[2]. Therefore, commercial vehicle fleets might play an important role in the future when it comes to sustainable mobility. On the one hand, the number of commercial car registrations is higher than that of private users (see Fig. 1.2). On the other hand, fleet managers are under pressure by internal emission reduction targets, increased costs for fossil fuels and a higher public attention for sustainability issues to procure vehicles with lower environmental impact.

However, the integration of alternative drives into a commercial fleet raises fundamental as well as operational questions: How much is the total energy demand of an electric car? How can the acceptance of electric cars by commercial users be increased? How can electric fleets be integrated into the existing electricity grid?

[2]Data origin: Kraftfahrtbundesamt: Neuzulassungen von Pkw im Jahr 2016 nach ausgewählten Kraftstoffarten, online verfügbar unter: https://www.kba.de/.

Under which conditions are electric vehicles beneficial from an environmental point of view? How can heterogeneous fleets be planned and operated?

These and other aspects are addressed by the joint research project Fleets Go Green. The project was funded from 2012 to 2016 by the Federal Ministry of Environment, Nature Conservation, Building and Nuclear Safety (BMUB) under the reference 16EM1041 and supervised by the Project Management Agency VDI/VDE-IT. It has been a joint effort by academia and industry. From academia four institutes of the Technische Universität Braunschweig participated, namely the Institute of Automotive Management and Industrial Production, the Institute of Automotive Engineering, the Institute of High Voltage Technology and Electrical Power Systems, and the Institute of Machine Tools and Production Technology, all associated through the Automotive Research Center Niedersachsen (NFF). Furthermore, the Fraunhofer Institute for Manufacturing Technology and Advanced Materials (IFAM) was involved. From industry several companies were engaged, including BS|Energy Braunschweiger Versorgungs-AG & Co. KG, imc Meßsysteme GmbH, ACTIA I+ME GmbH, iPoint-systems GmbH, Lautlos durch Deutschland GmbH, TLK-Thermo GmbH and Volkswagen AG.

The joint project aimed at the holistic analysis and assessment of the environmental performance of electric vehicles in everyday use by the example of corporate fleets. To this end, various battery-electric (BEV) vehicles were procured, equipped with appropriate measuring technology and operated in different application scenarios. The ecological and economic potentials were investigated and evaluated experimentally in fleet tests and theoretically with the help of component, vehicle and fleet simulations. The fleet tests enabled the recording, storage, preparation and analysis of diverse data for the vehicle, user and electricity network behavior in real operation. The simulation models enabled the modelling and evaluation of a large number of different drive train configurations with varying component properties as well as different fleet configurations.

An overview of the project structure is shown in Fig. 1.3. Within the five modules

- the real energy consumption of the fleet vehicles was measured, modelled and simulated (Module 1),
- the requirements of customers and operators were gathered in empirical studies, evaluated and transferred into suitable business models (Module 2),
- concepts for the increased integration of renewable energy sources into the electricity network were developed and implemented so that the environmental impact of the corporate fleet vehicles could be minimized (Module 3),
- a holistic assessment of the environmental performance (LCA) of electric and plug-in-hybrid vehicles in everyday use was carried out (Module 4),
- a simulation-based decision support for the ecologically-oriented fleet management has been developed so that an economic and environmental fleet operation could be implemented step by step (Module 5).

The main results of the project are documented in this publication. Within Fleets Go Green a sample of 25 vehicles from various manufacturers with a total mileage of more than 70,000 km and over 9300 individual journeys was analyzed (see Chap. 2). It turned out that the vehicles were mainly used in the inner city

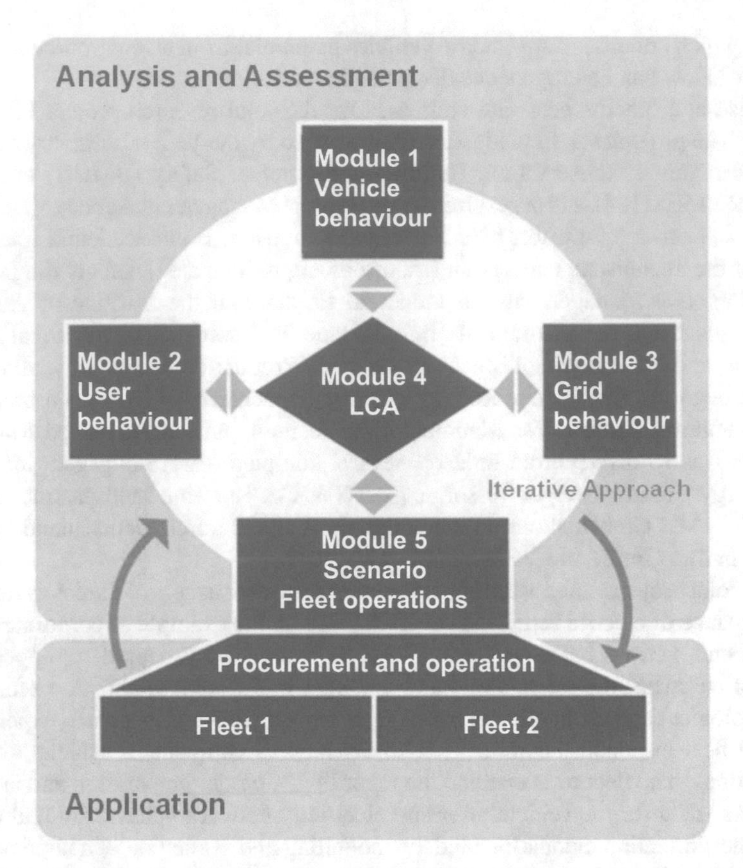

Fig. 1.3 Project structure of the joint research project Fleets Go Green

area and only short distances were travelled. Approximately 70% of all individual rides had a track length of less than 10 km and almost all journeys were shorter than 50 km. As a result, the range of battery-electric vehicles played hardly any role under the application conditions of the project.

The empirical studies (see Chap. 3) showed however that the range has a much greater influence on the purchase decision of a potential customer than the investigated rather short driving distances of the vehicles suggest. Thus, a reliable prognosis of the range of electric vehicles is an important criterion for using electric vehicles, especially in the case of weather conditions which demand air-conditioning. The demand for electrically operated vehicles exists across all vehicle sizes, although there is an increased willingness to buy plug-in hybrids. However, the factors that play a role in vehicle selection cannot be attributed to symbolism, hedonism or functionality, but rather to the enthusiasm and the environmental awareness of the interviewees.

The charging infrastructure also has significant influence (see Chap. 4). For example, the duration and the effort of the charging procedure have a significant influence on the purchase intention. Hence, the availability of the fleet should be increased by higher charging capacities. For the alternating current charge, the charging capacity of the onboard charging units of the vehicles should be increased. By means of targeted charging management, vehicles with higher charging capacities than the previous 3.7 kW can be easily integrated.

The life cycle assessment clearly showed that the environmental impacts of electric fleets strongly depend on the energy mix (see Chap. 5). Consequently, the share of renewable energies in the charging current must be increased. At the same time, the total energy consumption of the fleet in the utilization phase must be reduced. Here, external temperatures have a decisive effect on the overall energy consumption, since this is dependent on auxiliary consumers such as heating and air-conditioning. The respective climate zone in which the fleet is operated therefore also plays a central role in the decision as to which electric vehicles can be put to good use from the ecological point of view. The same applies to further geographic features such as route profiles and typical routes. Finally, also the driving behaviour of the users has a noticeable influence. Here, savings can be achieved through an appropriate training of drivers.

With regard to the developed decision support for fleet planning, general statements on the advantageousness of electric vehicles for fleet operations are not possible due to the high heterogeneity of fleets (see Chap. 6). Instead, the mapping, modeling and simulation of the specific decision situation of fleet operators is appropriate. Based on the Fleets Go Green application scenarios, it was shown that the integration of electric vehicles into corporate fleets can have positive effects with regard to the environmental impact associated with the fleets.

Additional to the detailed scientific results presented in Chaps. 2–6, the main findings from the project have been summed up as recommendations for actions (see the executive summary). With its holistic approach to analyze the vehicle, user and (energy supply) network behavior, the joint project "Fleets Go Green" provides the necessary insight for the exploitation of the ecological potential of electric vehicles in fleet operations.

Chapter 2
Fleet Measurement and Full Vehicle Simulation

Marcel Sander, Michael Gröninger, Ole Roesky, Michael Bodmann and Ferit Küçükay

2.1 Introduction

In Fleets Go Green electric vehicles were integrated in two vehicle fleets. On the one hand the factory fleet of BS|Energy and on the other hand a car-sharing pool concept for the Technische Universität Braunschweig. Both vehicle fleets consist of different types of electric vehicles from various manufactures. The vehicles are from the A and M segment according to the classification of the EU. Their parameters are shown in Table 2.1.

In total 13 Smart ED, ten VW e-up!, two Citroën C-Zero and one Nissan eNV200 were operated in the two fleets.

To gain knowledge about the usage profiles and the energy consumption as well as to verify the simulation most vehicles were equipped with measurement equipment. This equipment consisted mainly of a data logger, which recorded data from the CAN bus of the vehicle. Relevant CAN bus data had to be identified, since the manufactures of most vehicles did not provide the necessary information, due to the fact that they did not participated in Fleets Go Green. For this and to get additional data, three vehicles of different types were equipped with high voltage measurement equipment. These three vehicles were also used in regular fleet operation for two years. In addition to

M. Sander (✉) · F. Küçükay
Technische Universität Braunschweig, Institute of Automotive Engineering (IAE), Braunschweig, Germany
e-mail: marcel.sander@tu-braunschweig.de

M. Gröninger
Fraunhofer Institute for Manufacturing Technology and Advanced Materials (IFAM), Bremen, Germany

O. Roesky · M. Bodmann
TLK-Thermo GmbH, Braunschweig, Germany

C. Herrmann et al. (eds.), *Fleets Go Green*, Sustainable Production, Life Cycle Engineering and Management,
https://doi.org/10.1007/978-3-319-72724-0_2

Table 2.1 Parameters of the vehicles used in Fleets Go Green, based on (DAT 2015)

	Smart ED	Volkswagen e-up!	Citroën C-Zero	Nissan eNV200
Curb weight (kg)	900	1200	1195	2220
Peak power (kW)	55	60	49	80
NEDC energy consumption[a] (kWh/100 km)	15.1	11.7	13.5	16.5
EU segment	A	A	A	M
Seats	2	4	4	5
Use case	BS|Energy		TU Braunschweig	

[a]including the charging losses

the high voltage measurements there were also thermal measurements of the vehicle cabin in summer and winter.

Every type of electric vehicle had to be represented by the full vehicle simulation. Favourable for the simulation was, that all vehicles had the same topology. They had one electric machine (EM), with a single-speed gearbox and one drive axle (front or rear). Due to this it was possible to use the same structure for all vehicles and use different sets of parameters.

Focus of the simulation was to evaluate a realistic energy consumption in various scenarios and under different conditions, like ambient temperature, driving environment or vehicle loads. To achieve this, it was important to model the powertrain and the HVAC system (Heating, Ventilation and Air Conditioning) system. The HVAC system is the most significant of the auxiliaries and has a great influence on the overall energy consumption. Due to this the simulation should also be able to model different measurements to reduce the auxiliary energy consumption.

Besides the effects of this measurements, the simulation was also used for studies on the powertrain configuration, like the influence of the transmission ratio, power of the electric motor or capacity of the battery. The results of these studies were the basis for the fleet simulation.

2.2 Fleet Measurement

The focus of the fleet measurement was on the energy consumption and usage profile. The measurement took place from mid 2014 to mid 2016 and a total of 21 vehicles were equipped with measurement equipment.

Basis of the measurement equipment were data loggers which recorded data of the vehicles CAN bus. They were also equipped with GPS receivers to detect the position of the vehicle during trips. One trip started as the ignition was switched on and ended when the ignition was turned off. After one trip had ended the measurement

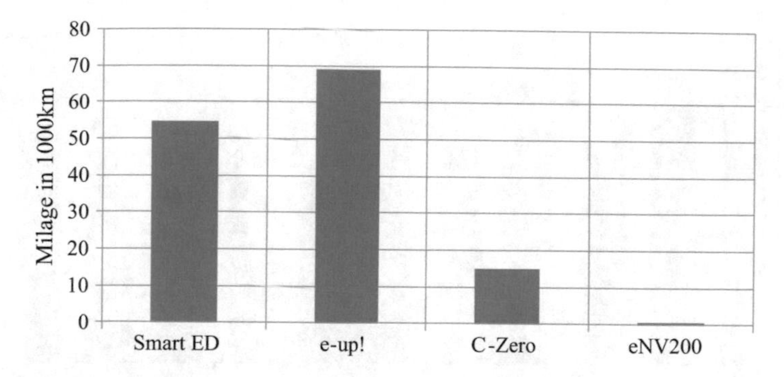

Fig. 2.1 Mileage with measurement data for each vehicle type

was finished. After that the data was transferred to a FTP server via a Wi-Fi access point at their parking position. Finally, a plausibility check was performed and some key parameters, like travelled distance and energy consumption, were calculated before they were transferred into a database. This database was the data source for all partners in Fleets Go Green.

Since most manufacturers of the vehicles did not participate in the project the relevant data on the CAN bus had to be identified. To achieve this and also to get additional data three vehicles one vehicle of each type (Smart ED, VW e-up! and Citroën C-Zero) was equipped with extensive high voltage measurement equipment. It was integrated into the high and low voltage grid to measure currents and voltages of the different components, especially of the heater, refrigerant compressor, high and low voltage battery, DC/DC converter and electric machine. With the measured data the related signals on the CAN bus could be identified.

Furthermore, thermal measurements of the passenger cabin were performed. For this temperature sensors were installed in the cabin of the vehicles and the temperature curves for heating and cooling were recorded. They were used to validate the parameterisation of the cabin model.

Figure 2.1 shows the mileage with measurement data available of each vehicle type.

2.3 Results of Fleet Measurement

Figures 2.2, 2.3 and 2.4 show the mean energy consumption for each month of the Smart ED, VW e-up! and Citroën C-Zero. The overall energy consumption is further divided into the consumption of the powertrain and the auxiliaries.

The figures show, that the energy consumption in winter is significantly higher than in summer. This is the result of a much higher energy demand from the auxiliaries. The demand of the powertrain does not differ much between Smart ED and e-up! whereas the C-Zero requires less energy. A reason for this could be a different vehicle

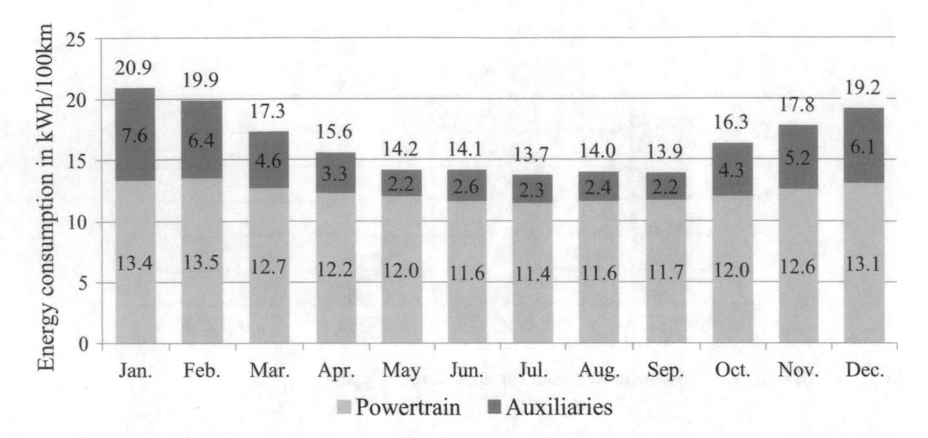

Fig. 2.2 Energy consumption of the Smart ED for each month divided into powertrain and auxiliary energy consumption

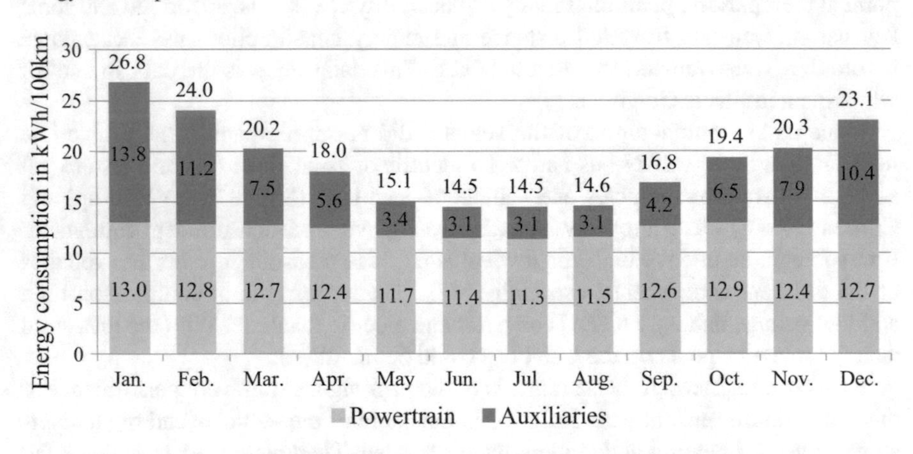

Fig. 2.3 Energy consumption of the VW e-up! for each month divided into powertrain and auxiliary energy consumption

usage regarding acceleration and vehicle speed. Moreover, differences between the auxiliaries are present. The Smart ED has a lower energy consumption due to the smaller size of the passenger cabin.

A comparison of the mean energy consumption of all four vehicle types used in Fleets Go Green shows Fig. 2.5.

The main differences could also be explained with the vehicle size. For this reason, the highest energy consumption has the biggest vehicle, the Nissan eNV200. But it can also be seen, that vehicles of similar size also differ. Besides trivial reasons for this, especially different vehicle parameters, like weight or drag coefficient, and different drivers with different driving styles, there is another significant factor: the use of the vehicles during a year. Figure 2.6 shows the share of mileage per month of the Smart ED, e-up! and C-Zero.

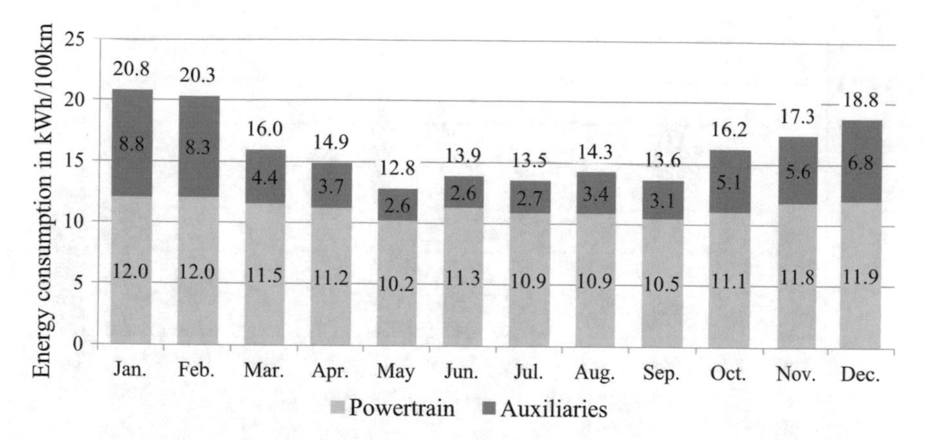

Fig. 2.4 Energy consumption of the Citroën C-Zero for each month divided into powertrain and auxiliary energy consumption

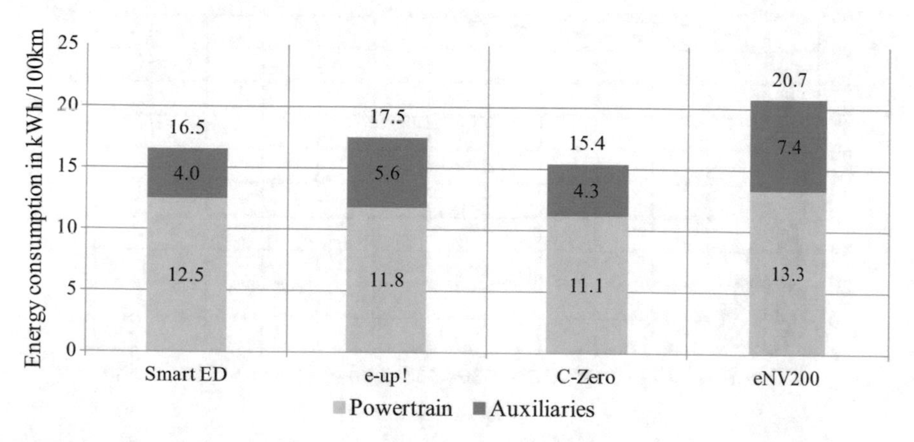

Fig. 2.5 Average energy consumption of the four vehicle types divided into powertrain and auxiliary energy consumption

The distribution of mileage per month of the Smart ED and e-up! is similar. An exception is the C-Zero which has a considerable less mileage in winter. This has a great influence on the auxiliary energy consumption, since the heating in winter requires more energy than cooling in summer. Due to this the overall energy consumption of the auxiliaries of the C-Zero is lower compared to the e-up! although both are comparable in regard of the cabin size.

Regarding the usage profile especially the travelled time and distance were analysed. The result is, that the trips were short in both, time and distance. But there were also some differences between the fleet of BS|Energy and the TU Braunschweig.

90% of the trips of the commercial fleet of BS|Energy were shorter than 15 min and 5 km. These extremely short distances are the result of the use case at BS|Energy on the one hand and the definition of one trip on the other hand. A tour was planned

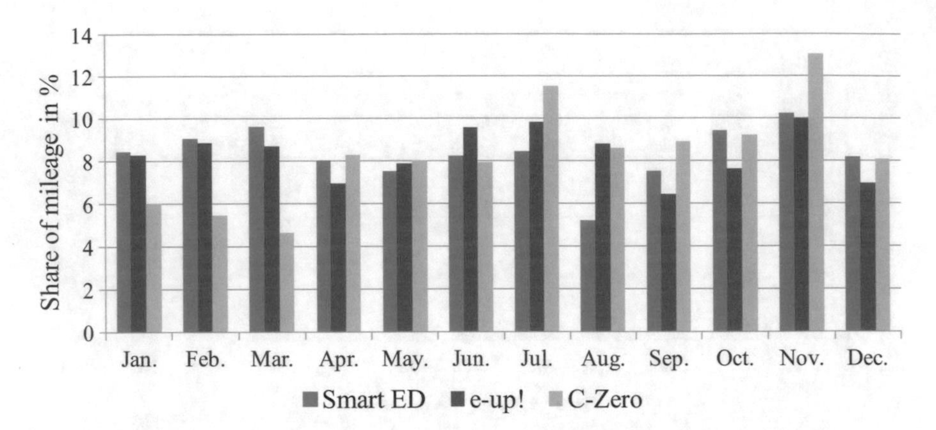

Fig. 2.6 Share of mileage per month for the Smart ED, e-up! and C-Zero

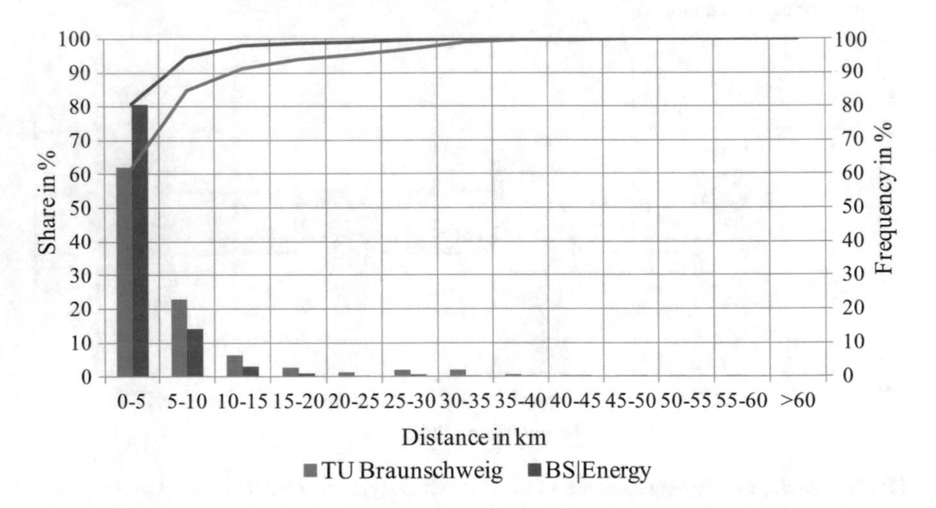

Fig. 2.7 Share and frequency of the travelled distances per trip for BS|Energy and TU Braunschweig

in a way, that the customers to visit are very close to each other, often just a couple of hundred meters between two stops. The frequency of travelled time and distance for BS|Energy are shown in Figs. 2.7 and 2.8.

The TU Braunschweig pool fleet vehicles also travel short distances, but the vehicles are also frequently used for trips in surrounding cities, especially Wolfsburg. Due to this there is a higher frequency of longer distances and travelled time. The result is that in this use case 90% of the trips are shorter than 20 min and 10 km. The frequencies of the TU Braunschweig are also shown in Figs. 2.7 and 2.8.

The mentioned differences in the usage of the both fleets are also reflected in the share of the driving environment. This is illustrated in Fig. 2.9, which shows the share of the mileage on urban, extra urban and highway roads. While both fleets have

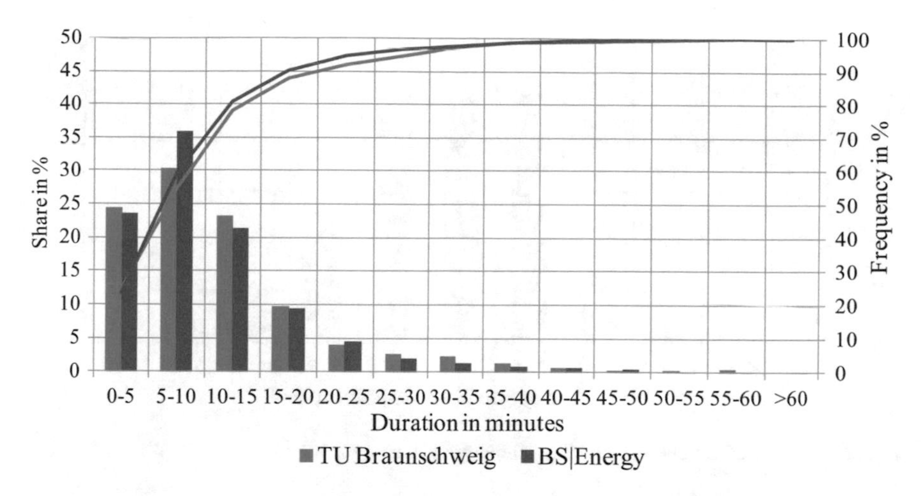

Fig. 2.8 Share and frequency of the duration of trips for BS|Energy and TU Braunschweig

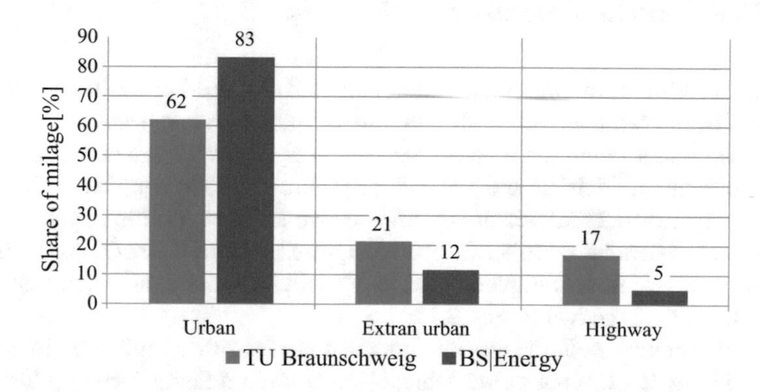

Fig. 2.9 Share of mileage on different road types for BS|Energy and TU Braunschweig

the highest share on urban roads, the TU Braunschweig vehicles have a significantly higher share on extra urban and highway roads than the ones of BS|Energy.

Another aspect of the driving environment is the ambient temperature. In Fig. 2.10 the average temperature during one trip is illustrated.

Due to the definition of one trip there are also very low (< −5 °C) and high (> 25 °C) mean temperatures. During most of the trips the average ambient temperature was between 0 and 20 °C.

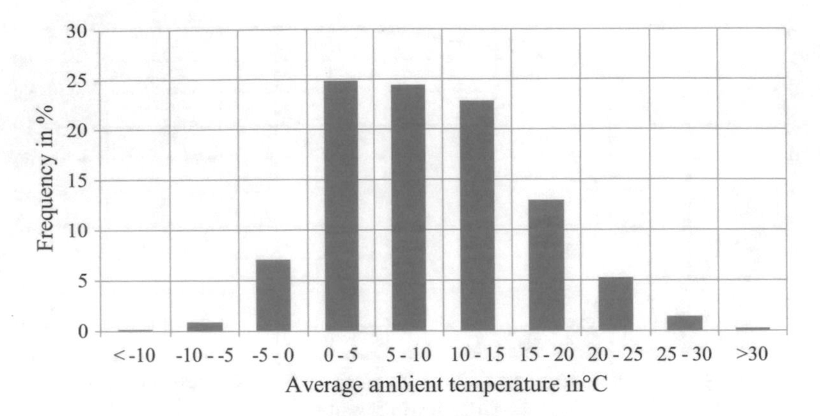

Fig. 2.10 Share of the average ambient temperature during the trips

2.4 Full Vehicle Simulation

Basis of the vehicle simulation are different models for the drivetrain, the electrical components, the cabin as well as the cooling, refrigeration and heating cycle. These models take into account the mechanical, electrical and thermal behaviour.

The different models of the full vehicle simulation are coupled with the co-simulation platform TISC which is a middleware to allow for the continuous synchronization of data between a wide range of different simulators (Kossel 2011). In the present set-up models implemented in Modelica/Dymola and Matlab/Simulink are coupled. This is shown in Fig. 2.11.

The full vehicle simulation was implemented as a forward simulation. Its structure is shown in Fig. 2.12. Input of the vehicle simulation is a target speed profile. Form the difference between the target and current speed the driver model determines an accelerator and brake pedal position. As an alternative a statistical driver model of the IAE can be used, which reproduces the measured vehicle usage of the drivers (speed, acceleration etc.) based on statistics. This model as well provides the pedal positions as an output. Based on this the model of the power electronics and the electric machine determines the EM torque, which is the input for the transmission model. Here the losses and gear ratio are modelled and the torque of the wheels is calculated. Out of this torque the traction force is determined, which is used to calculate the speeds of the vehicle, wheels and motor.

The electric machine as well as the heating and air conditioning system are the main consumers in the electric cycle. These high voltage and the low voltage components determine the current of the battery. From this current requirement the battery model calculates the voltage.

To model the energy consumption of the auxiliaries a single zone model of the passenger cabin is used. As the main components of the auxiliaries the HVAC system with a PTC (positive temperature coefficient) heater, a refrigerant compressor and a heat pump using R744 (CO_2) as a refrigerant are modelled.

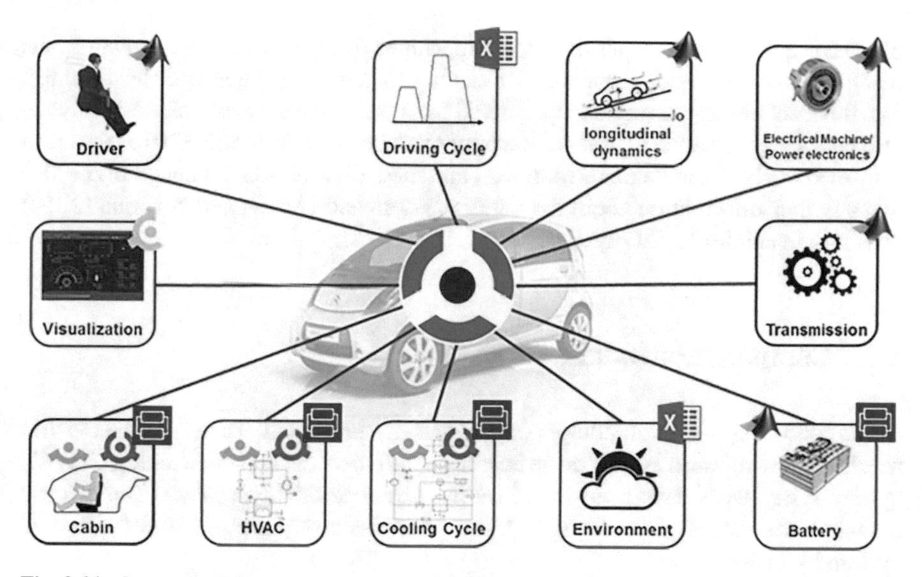

Fig. 2.11 Coupling of the various simulation models with TISC

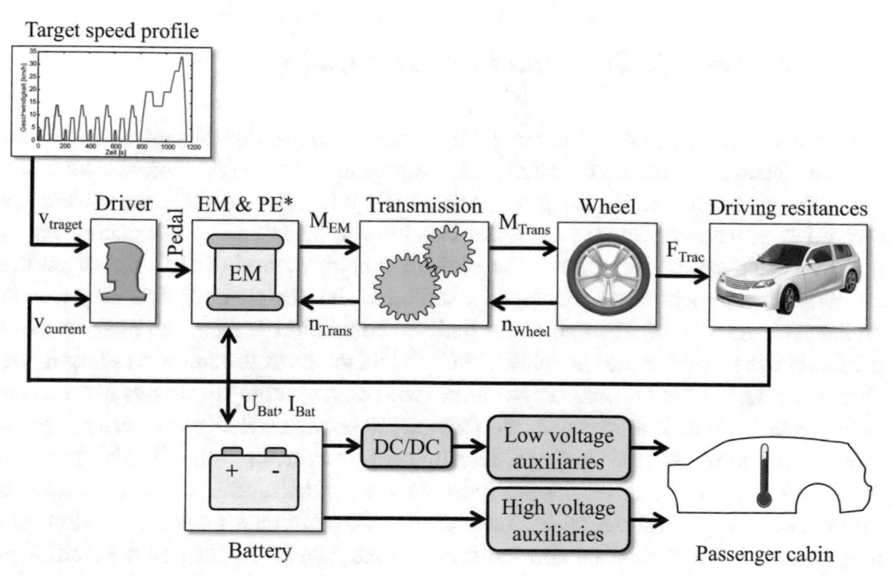

Fig. 2.12 Structure of the full vehicle simulation

The approach of simulation models for each component with defined interfaces creates a modular full vehicle simulation. To simulate the different vehicle types only their specific parameterization and knowledge about their component setup is needed. This enables the evaluation of different vehicle configurations. For example, the use of different types of electric machines (permanent magnet synchronous machine or asynchronous machine) or heating systems (PTC heater or heat pump). The different model parameters are derived from own measurements as well as literature data. Regarding the Citroën C-Zero for example, Eckstein et al. (2013) provides information about the vehicles road load and mass distribution, Kitano et al. (2008) gives further information about the battery system and Umezu and Noyama (2010) details the vehicles HVAC system.

2.5 Component Models

In the following, the main component models are presented. These are the electric machine, transmission as well as the heating, ventilation and air conditioning (HVAC) system. They were developed based on the expertise of the partners. The electric machine was modelled by the IFAM, the transmission by the IAE and the HVAC system by TLK-Thermo.

2.5.1 Electric Machine and Converter Model

All vehicles investigated in this project have center drives. For the component models, we selected the most commonly used e-machine types-induction machine and synchronous machine. Since the e-machine is by far the vehicle's largest energy consumer, mechanical input and power loss have to be determined as accurately as possible. This can be done using the Finite Element Method (FEM). Based on the induction and synchronous machines installed in the vehicles, the first step was to create with the help of FEM software realistic simulation models and analyse them for the entire speed/torque range. Besides the I^2R losses of the stator winding in the electric cycle, it was thus possible to predict with high accuracy the losses in the magnetic cycle, such as iron losses in the ferromagnetic material or eddy current losses induced in the permanent magnets. During post-processing with a Matlab routine, operating maps are calculated from these data, considering the respective machine limits like voltage, torque, speed and current. Through data parameterization, the map calculation routine considers additional losses caused by the power electronics that is needed for operation. For this reason, operating maps calculated once can be scaled to other efficiency classes of the same machine type. Resulting inaccuracies, which cannot be avoided, are assumed to be negligible here. Figure 2.13 shows a selection of operating maps for a PMSM.

A transient temperature model in Matlab/Simulink is the interface to other component models and to TISC. By k_r factor feedback, it considers additional self-heating caused by internal losses in individual heat zones of the electric machine, as for example in end-winding, tooth, rotor and housing. This means that temperature rise

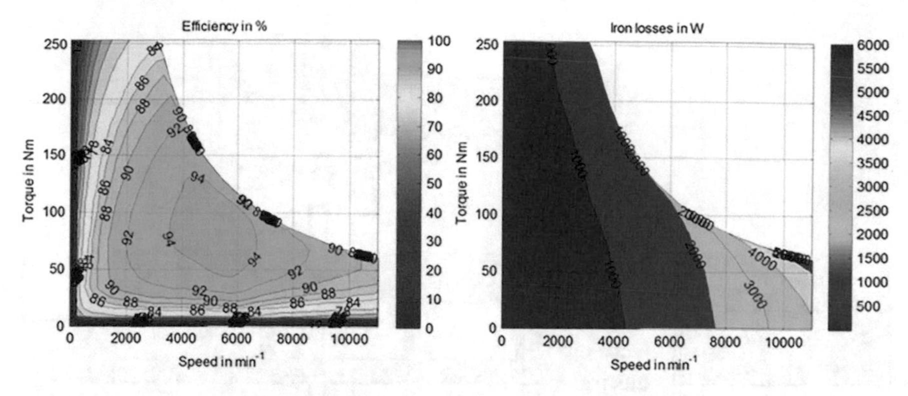

Fig. 2.13 Selected PMSM operating maps

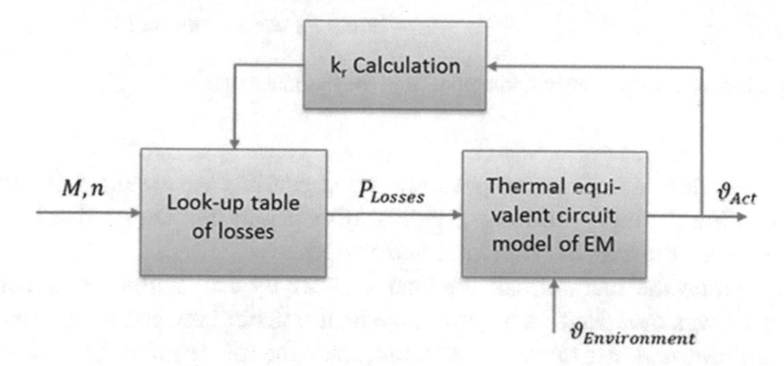

Fig. 2.14 Flow chart of the thermal component model for an e-machine

in the electric machine is subject to time-dependent load cycling, as for example defined by the NEDC. Figure 2.14 shows a simplified flow chart.

2.5.2 Transmission Model

Basis of the transmission model are characteristic diagrams, which contain the drag torque. The drag torque is determined depending on speed and torque at the transmission input as well as the temperature of the oil sump. With the temperature as an input it is possible to take the thermal loss behaviour of the transmission into account.

Due to the different transmission ratios there are specific characteristic diagrams for each vehicle type. They were generated with a validated transmission loss calculation program which was developed at the IAE. This program takes into account the geometric structure, like centre distance as well as position of the gear wheels and bearings, losses caused by bearings, toothing, seals and the interaction of the gear wheels with the transmission oil (Inderwisch 2015; Seidel and Küçükay 2015).

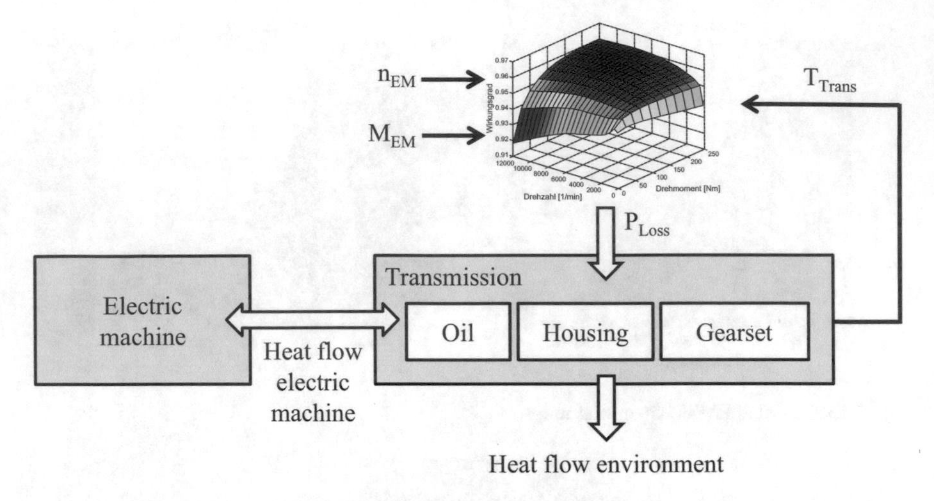

Fig. 2.15 Structure of the transmission model and the thermal balance

Since all vehicles have a two-stage single-speed gearbox we assumed that they are equal regarding geometry, bearings, sealing, dimensions etc. Due to that in the loss calculation only the transmission ratio was varied.

To determine the thermal balance heat transfer by convection, conduction and transmission was modelled. In addition, the heat transfer between transmission and electric machine was also taken into account, since the rotor shaft of the motor is also the input shaft of the gearbox. To calculate the gearbox temperature from the thermal balance it was assumed, that the transmission consists of three thermal masses: housing, oil and gear set. These three masses differ regarding material (different specific properties) and weight. Their parameters were chosen based on in detail analysed single-speed transmissions at the IAE and were also assumed as similar for all vehicles.

Figure 2.15 shows the structure of the transmission model.

2.5.3 HVAC System Model

The entire HVAC (Heating, Ventilation and Air Conditioning) system is implemented in Modelica/Dymola using the model library TIL jointly developed at TLK-Thermo and the Institute of Thermodynamics at the Technische Universität Braunschweig. Two different systems, a PTC (resistive heating) and a heat-pump using R744 CO_2, have been implemented to compare different heating and cooling strategies.

The conventional HVAC system with a PTC, the refrigeration cycle, the cooling cycle, the air path to the cabin is shown in Fig. 2.16. The refrigeration cycle consists of a compressor, an ambient heat exchanger (condenser), a receiver, an expansion valve and an evaporator. The refrigerants and coolants flow through the components

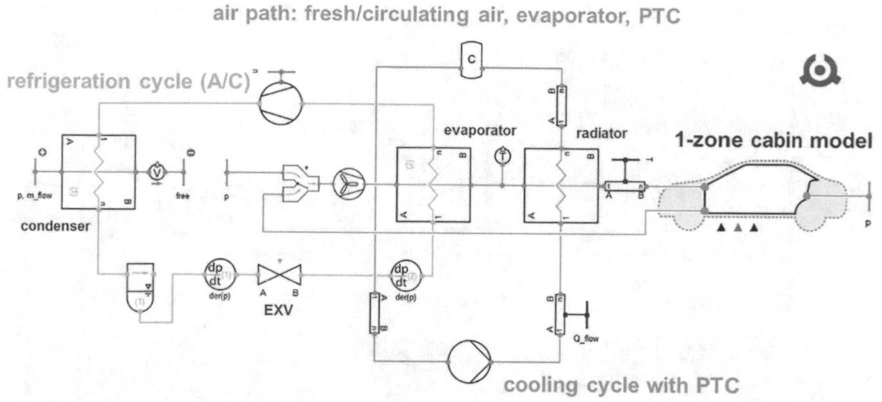

Fig. 2.16 HVAC system using PTC (overall system in TIL: PTC, refrigeration cycle and cabin)

in the same order they are enumerated. The cooling cycle consists of a pump, the PTC, a heat exchanger and an expansion tank. The airflow to the cabin model first passes the fresh/circulating air damper, the fan, the refrigeration cycle's evaporator, the cooling cycle's gas cooler and the air ducts within the vehicle's dashboard before entering the cabin.

Figure 2.17 shows the HVAC system when using a CO_2 heat pump to provide heating to the vehicle cabin. The refrigerant in the heat pump system flows through the components in the same order they are enumerated below. It consists of a compressor, a gas cooler, the first expansion valve, frontend heat exchanger, a second expansion valve, a 2/3-way valve, the evaporator and a receiver. By means of the 2/3-way valve different operating modes can be used. When heating the cabin, the 2/3 way valve is set for the refrigerant to bypass the evaporator and only pass through the gas cooler and the frontend heat exchanger to harvest heat from the outside and provide heat to the cabin. When using the system for air conditioning, the refrigerant flows through every component. In this case the airflow towards the cabin bypasses the gas cooler to only provide low temperature air to the cabin. Except for the gas cooler bypass, the air path functions in the same way previously described for the conventional HVAC system.

2.6 Validation

In the following section the validation of the full vehicle simulation and the EM model are shown.

The full vehicle simulation is validated by using the speed profile as well as the conditions of the environment of a measured trip as the input for the simulation and comparing the results of the simulation with the measurement. For the EM model an

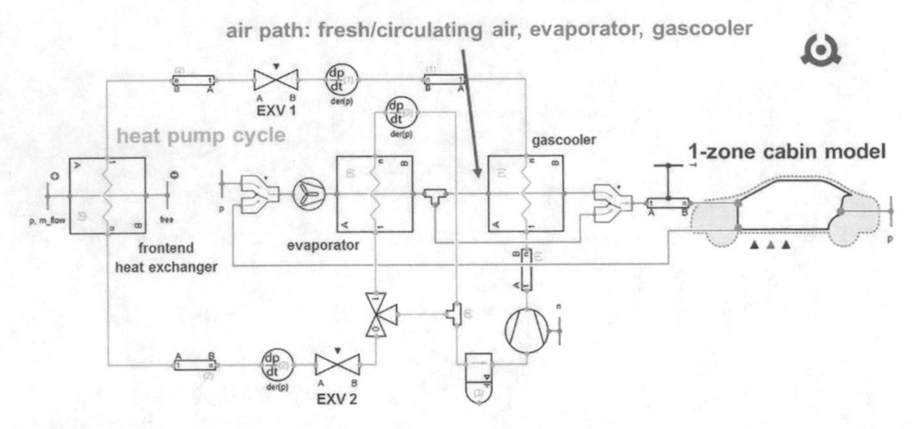

Fig. 2.17 HVAC system using CO_2 heat pump

electric machine (PMSM) was tested on a test bench of Fraunhofer IFAM and the results were compared with the simulation as well.

2.6.1 Validation of Full Vehicle Simulation with Real Driving Profiles

As mentioned before the recorded data was also used to validate the simulation model. This is shown in the following using an example of a trip of a C-Zero. The route of the trip is shown in Fig. 2.18. It took place in October 2014, started in Braunschweig and the destination was one of the surrounding villages (Essehof). The ambient temperature at that time was approximately 17 °C. In Fig. 2.18 the speed and height profile is also shown.

The measured speed as well as the environment conditions, like the ambient temperature, were used as the input of the simulation. Figure 2.19 illustrates the result of the simulation and contains the speed profile as well as the simulated and measured energy consumption.

It is visible, that the consumed energy of the simulation and the measurement are very close together. This proves, that the simulation delivers satisfying results.

The validation was performed for all vehicles at different environment conditions. To summarize, all evaluations show that the simulation is able to determine the energy consumption with sufficient precision for various situations, like different environment conditions, vehicle configurations etc.

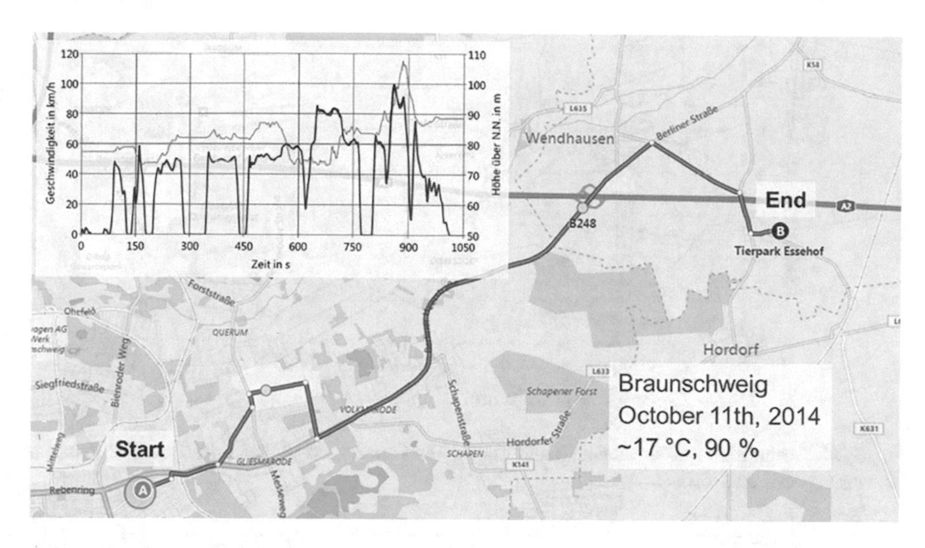

Fig. 2.18 Route as well as speed and height profile of the trip used to validate the full vehicle simulation (Jugert and Fischer 2014)

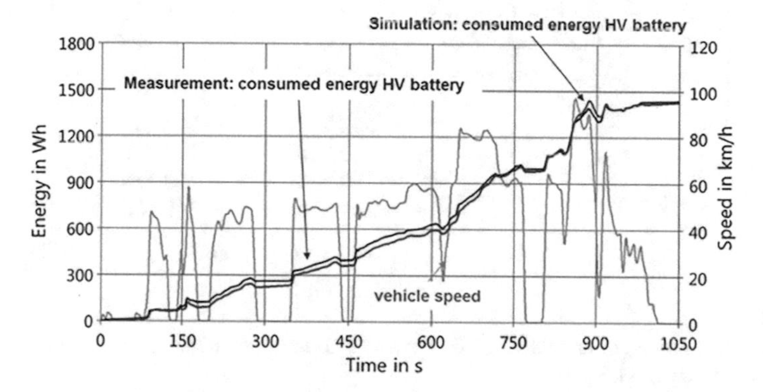

Fig. 2.19 Speed profile and consumed energy of simulation and measurement (Jugert and Fischer 2014)

2.6.2 *Validation of the EM Model with Test Bench Measurements*

To validate the drive train's component models, Fraunhofer IFAM was provided with a PMSM including converter. The electric machine was installed and tested at the institute's performance test bench. Selected measurements are depicted in the following figures, comparing the results of the data derived from FE simulations with the test bench measurements made at four different torques up to 8000 min^{-1}.

The measured values show good agreement with the simulated values (Figs. 2.20 and 2.21), only deviating at maximum torque of 200 Nm. The reason might be that

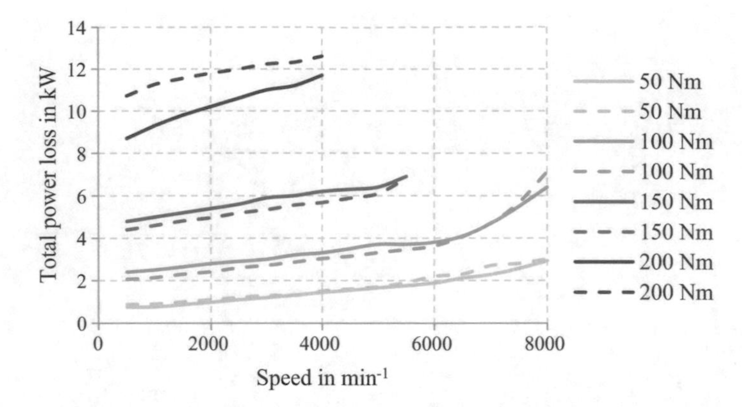

Fig. 2.20 Comparison of simulated (solid curves) and measured (dashed curves) total losses at different speeds and torques

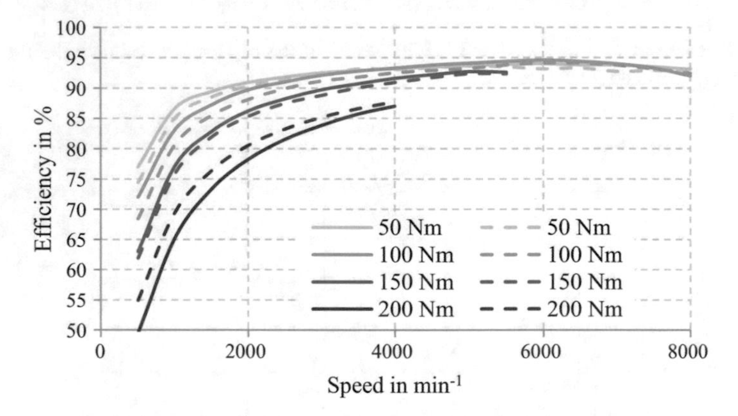

Fig. 2.21 Comparison of simulated (solid curves) and measured (dashed curves) total efficiency at different speeds and torques

in case of very high currents, the real converter has lower losses than the simulated one which, for lack of detailed converter data, was modelled based on empirical data. However, considering the evaluated data, it can be assumed that the simulated converter was modelled with sufficient accuracy. Since the deviations between simulation and test values are negligibly small, the validation is considered as successful.

2.7 Simulation Studies

Using the validated full vehicle simulation described in the previous section, different simulation studies have been conducted. Simulation studies comparing different

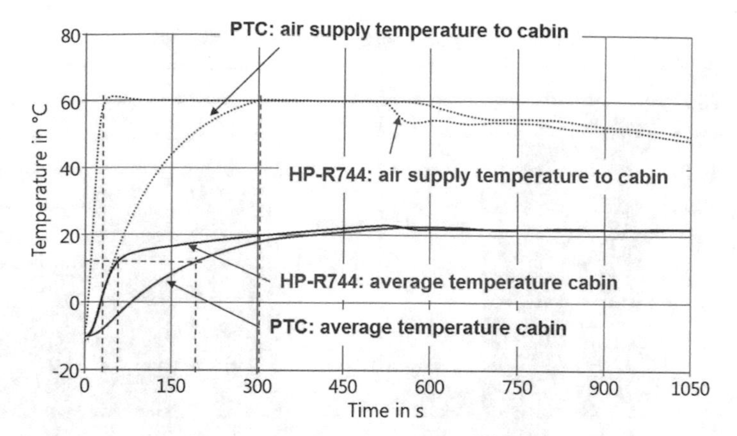

Fig. 2.22 Average temperature of the cabin and the air supply temperature to the cabin for PTC and heat pump-system at −10 °C ambient temperature (Jugert and Fischer 2014)

HVAC systems to heat the vehicle's cabin will be detailed in this section. Due to the efficient powertrain of electric vehicles there is relatively few waste heat to harvest in order to heat the cabin. The energy used to heating, either by means of a PTC or a heat-pump system, is typically drawn from the high voltage battery. In winter months high heating requirements may drastically reduce the vehicle's driving range. Therefore, different HVAC systems have been investigated in terms of heating capacity and efficiency. A conventional PTC heating system is compared to a heat pump system using R744 CO_2 as a refrigerant. The detailed model description can be found in the section component models. Different simulations at different ambient temperatures have been conducted (−15, −10, 0 and 20 °C as a reference).

In Fig. 2.22 the average temperature of the cabin as well as the air supply temperature provided to the cabin are displayed for both the PTC as well as for the heat pump system at an ambient temperature of −10 °C.

It can be seen that the heat pump system requires only around 30 s to provide the desired air supply temperature of 60 °C whereas the PTC system requires 300 s. This results in a much faster heat up of the cabin temperature. Using a heat pump system, the cabin temperature reaches 12 °C after approximately 60 s whereas using a PTC it takes approximately 210 s to reach the same cabin temperature.

The results of the different simulations at different ambient temperatures are compared in Fig. 2.23. A simulation at 20 °C ambient temperature without heating or air conditioning is used as a reference to compare the increase of energy consumption when using a PTC or heat pump system to heat the cabin at −15, −10 and 0 °C. The energy consumed using the heat pump system is significantly lower than using a PTC under every considered condition. At the same time more heating energy is provided to the cabin which results in a higher passenger comfort. When using a heat pump system the increased energy consumption can be reduced from 34% at −15 °C up to 51% at 0 °C compared to using a PTC.

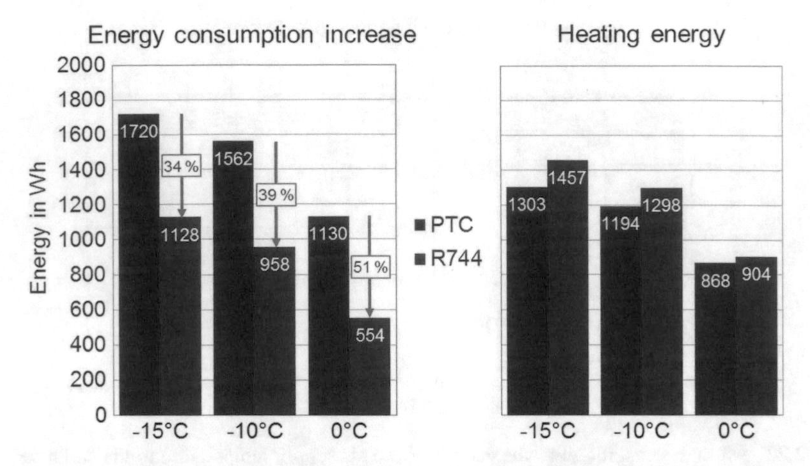

Fig. 2.23 Energy consumption increase (compared to 20 °C reference) and heating energy provided to cabin at different ambient temperatures (Jugert and Fischer 2014)

2.8 Conclusion

The full vehicle simulation created within the research project fleets go green has been validated using measurements taken under real driving conditions. The simulation shows good agreement with the measurements under different environment conditions. Different simulation studies have been undertaken. For the studies regarding the HVAC system, different components have been developed and a comparison of a conventional system using a PTC and a heat pump system using R744 CO_2 as a refrigerant has been conducted.

References

Deutsche Automobil Treuhand GmbH (DAT) (2015) Leitfaden über den Kraftstoffverbrauch, die CO_2-Emissionen und den Stromverbrauch aller neuen Personenkraftwagenmodelle, die in Deutschland zum Verkauf angeboten werden, Ostfildern

Eckstein L, Göbbels R, Wohlecker R (2013) Benchmarking des Elektrofahrzeugs Mitsubishi i-MiEV. In: Siebenpfeiffer W, Energieefiziente Antriebstechnologien (S. 22–29). Wiesbaden:Springer-Vieweg

Inderwisch K (2015) Verlustermittlung in Fahrzeuggetrieben; Dissertation, 2014; Schriftenreihe des Instituts für Fahrzeugtechnik, TU Braunschweig; Shaker, Aachen

Jugert R, Fischer H (2014) Fahrgastraumkonditionierung von Elektrofahrzeugen in Winterlastfällen; 3. VDI-Fachkonferenz Thermomanagement für elektromotorisch angetriebene PKW, Manching bei Ingolstadt, 5–6 Nov 2014

Kitano S, Nishiyama K, Toriyama J, Sonoda T (2008) Development of Large-sized Lithium-ion Cell "LEV50" and it's Battery Module "LEV50-4" for Electric Vehicle, GS Yuasa Corporation

Kossel R (2011) Hybride Simulation thermischer Systeme am Beispiel eines Reisebusses; Dissertation

Seidel T, Küçükay F (2015) Practice-oriented determination of transmission efficiency. CTI Symposium USA, Novi, USA

Umezu K, Noyama H (2010) Air-Conditioning system for Electric Vehicles (i-MiEV). SAE Automotive Refrigerant & System Efficiency Symposium 2010, Scottsdale Arizona:SAE International

Chapter 3
Determining Relevant Factors in Purchasing Electric Vehicles for Fleets

Daniela Mau and David M. Woisetschläger

3.1 Introduction

Globalization and technological progress have increased the demand in mobility. When companies operate, their products and employees need to be mobile as well. While these mobility needs come with further demand in energy to bring products and people from one place to the other, the question remains how the use of natural resources and CO_2-emissions can be reduced.

One way to do so is by electrifying mobility and using renewable energy, as the traffic segment is responsible for a large percentage of national CO_2 emissions. For example, the traffic sector in Germany emitted almost 18% of all CO_2-emissions in 2015, which is 8% more than in 1990 (German Federal Environmental Agency 2016). Especially local transport can be handled well with electric vehicles (EV), as on the one hand, their daily driven kilometers fall within the range of EVs, and on the other hand, they are often parked at the business location overnight, where they can be recharged. Governments worldwide have therefore introduced programs to incentivize the purchase and use of EVs. For example, California (USA) introduced a zero-emission-policy forcing car manufacturers to develop more sustainable friendly cars (Sierzchula et al. 2012); the car manufacturer Tesla being one example. Norway subsidized EV purchases while at the same time introducing charging infrastructure, thereby increasing the percentage of EVs of all cars in Norway significantly (Holtsmark and Skonhoft 2014). Germany funded several showcase projects with about 180 million Euro to increase research and use of EVs (Showcase Electric Mobility Germany 2014).

Company car fleets can be seen as pacemaker in the diffusion of EVs, as they own the largest percentage of cars (Nesbitt and Sperling 2001). In Germany, 63.8%

D. Mau · D. M. Woisetschläger (✉)
Chair of Services Management, Technische Universität Braunschweig, Braunschweig, Germany
e-mail: d.woisetschlaeger@tu-braunschweig.de

C. Herrmann et al. (eds.), *Fleets Go Green*, Sustainable Production, Life Cycle Engineering and Management,
https://doi.org/10.1007/978-3-319-72724-0_3

of all new registered cars were owned by companies (KBA 2016). The usage frequency of commercial cars is higher than for privately owned ones. Due to the lower consumption costs, the break-even point where EVs are similarly expensive than internal combustion engine vehicles (ICEVs) is reached sooner. Moreover, EVs are often driven back to company grounds, which suits well for the implementation of charging infrastructure (Nesbitt and Sperling 2001: 297f).

Literature on the diffusion of EVs has mainly focused on private households, while research on fleets is rare (Kaplan et al. 2016). This chapter therefore attempts to contribute to the understanding of fleet vehicle purchasing. In particular barriers and drivers of organizational EV adoption are identified and their impact on EV purchase intention are examined. While research has largely focused on Total Costs of Ownership (TCO) and driving range of EVs, it is a point of interest to determine the relative influence of these factors on purchase intention. An answer to this question sheds light on the question what to improve first in order to enhance the diffusion rate of EVs.

In accordance with these research objectives, the chapter is organized as follows. In the next section, we briefly acknowledge existing research on organizational fleet purchase behavior and the adoption of electric vehicles in fleets. Due to the sparse literature on organizational EV adoption, a preliminary qualitative study has been conducted. Building on the findings of the preliminary study as well as existing literature, we develop a theoretical framework for the proposed effects of range and acquisition price on purchase intention by fleet managers. Study 2 tests this relationship and furthermore demonstrates the impacts of TCO, mobility needs and image fit on purchase intentions of fleet vehicles. Results indicate that either the price or the range have to be matched to the level of conventional cars in order to achieve a similar level of purchase intention. In the last section of the paper we discuss implications for research and organizations.

3.2 Literature Review

Literature on fleet vehicle purchasing is sparse (Nesbitt and Sperling 2001; Kaplan et al. 2016) and only few studies explore motives and barriers for EV fleet adoption (Kaplan et al. 2016; Müller et al. 2015). Several studies looked at the drivers and barriers of private households. Here, the main barriers to EV adoption are found to be financial barriers (high purchase costs) and technological barriers leading to functional limitations (limited range and long recharging time), whereas drivers were recognized to be, amongst others, low running expenses, reduced maintenance costs (Hindrue et al. 2011) as well as psychological factors such as attitudes and lifestyles (Lane and Potter 2007). While some barriers and drivers are probably similar in their influence on purchase intention, the relationship is still unclear and needs further research.

However, the organizational context has largely been neglected by researchers, with some notable exceptions. Many studies emphasized the suitability of electric vehicles in fleets, mainly due to their plannability of routes, financial power and the possibility to switch only parts of their fleets so that longer-distance routes can be driven by internal combustion engine vehicles (ICEVs) (Kley et al. 2011; Gnann et al. 2015). In a qualitative study with fleet managers (Sierzchula 2014) found that drawbacks of EVs in fleets are especially related to negative TCO, the lost time during charging, the resulting lack of operational capabilities and the inaccuracy of driving range information by manufacturers.

Especially TCO has been given more attention in research (Hagman et al. 2016). TCO for electric vehicles and ICEVs are considered to be equally important for fleet managers (Contestable et al. 2011). The drawback here is that electric vehicles become only profitable when driven longer distances (Wu et al. 2015). An integration of electric vehicles is also dependent on the fleet size (Doll et al. 2011), as enough ICEVs must be present to substitute electric vehicles for trips that have longer distances (Hiermann et al. 2016). The upside of EVs in fleets is demonstrating to participate in climate protection, thereby enhancing their image (Grausam et al. 2015). Even though literature on EV adoption in general points towards acquisition price and driving range as the major determinants of purchase intention, sources on drivers and barriers are still few, so that we conducted a preliminary, qualitative study to check whether these are indeed the only motives in fleet purchasing besides recharging time.

3.3 Preliminary Qualitative Study

The aim of the preliminary study is to shed light in the factors influencing fleet purchase in general and for EV in particular. As research on EV adoption in fleets is sparse, a semi-replicative study to the one of (Sierzchula 2014) is performed in order to see whether there are more factors influencing decision-making of fleet vehicles in general and EVs in particular. As (Sierzchula 2014), qualitative, semi-structured interviews have been conducted with fleet managers. However, we included also car manufacturers and dealers into our sample to get a more general view of the decision-making process from a larger perspective than the one of a single company.[1]

In total 39 experts from 35 different companies were interviewed during 2013. The average duration length was 30 min. Respondents were mainly responsible for fleet management and general management in the automotive and manufacturing industry. Most of the companies interviewed had more than 250 employees. Thirteen of the companies have already included EVs in their fleets.

The interviewees were asked to first introduce themselves and their task in the fleet purchasing, continued by their procedure in buying new vehicles and the factors influencing their vehicle choices. Later they were also asked about the suitability of EVs in their fleet and drivers and barriers for their integration in fleets.

[1] An earlier version of the preliminary study has been published (cf. Müller et al. 2015).

Respondents mainly spoke about three aspects that are important in the decision-making for fleet vehicles, namely TCO, functional attributes and image aspects. The importance of these three aspects seems to change according to the purpose of the vehicle. While manager cars shall demonstrate prestige, pool cars are usually small and unobtrusive. Manager cars may cost more and may even emit more CO_2 than pool cars.

With regard to TCO of EVs, respondents unanimously said that current EVs are still too expensive to be competitive. Most of the respondents who already had EVs in their fleet adopted them within the framework of showcase projects.

> The cars are subsidized. We get money from the state for the cars and the infrastructure. (IP 17)

Even though some respondents were skeptical about the usefulness and environmental friendliness of EVs, the majority liked them.

> Somewhen, when the cars are capable of actually being used and if then the price is in an economically interesting region, it almost does not matter which car I drive. (IP 03)

Of major concern were functional attributes. These were especially related to the range whereas the long recharging duration was hardly discussed.

> "The range is the biggest problem for us. In general I like them, and as soon as the range and the price fit, I will definitely purchase one." (IP 25) "It is a matter of the price and of the functional value, that is: can I use the car in a way that I have to. Does the car really have those advantages and how do they relate to the economic value?" (IP 03)

> At the moment, there is a certain flavor that if I drive an electric vehicle, I do well for the environment, but I have to restrict myself somewhere else. (IP 12)

Image reasons were named to a lesser extent than the other two aspects. However, in the discussion on EVs in particular, the positive impact of EVs on the company image are mentioned more frequently.

> You invest in a future-oriented technology, which is not bad for the company image. Such things can always be shown. (IP 18)
>
> We try to outperform the required environmental standards (IP 06)

The aim of this preliminary study was to shed light on factors influencing fleet purchase decisions. It was found that there are three major factors, namely TCO, functionality and image. These findings go generally in line with the findings by (Sierzchula 2014) who also claimed that these factors are relevant. The situational factors of (Lane and Potter 2007) can also be found in our findings, even though they focused on users and not fleet managers. With regard to the findings of (Lieven et al. 2011) who put acquisition costs and driving range on a continuum, we added a third dimension, namely the image. Thus, we could replicate findings of (Sierzchula 2014) and (Lane and Potter 2007) for fleet managers. The question remains which factors influence the adoption decision for EVs of fleet managers the most and significantly.

3.4 Main Quantitative Study

3.4.1 Hypotheses Development

In the preliminary qualitative study we found that the major factors influencing fleet vehicle choice are TCO, functionality as well as image or symbolic related attributes. We like to validate these qualitative findings with a larger sample and determine whether these factors actually influence purchase intention. As image related factors are very different to manipulate as they are very subjective, we wanted to know which of the two other factors, namely acquisition costs and driving range as functional attribute, is the most important factor and whether an improvement in one factor substitute the drawbacks of the other. Therefore, a quantitative survey was conducted with fleet managers in Germany.

Our preliminary study found that EVs need to fit to the company in terms of TCO, functionality and image. TCO is mainly related to the acquisition price or respectively elevated leasing rate of EVs. On average, EVs are roughly 10,000 € more expensive than ICEVs with same brand, size and supplementary equipment. Companies lack the financial possibility due to cost pressures, which is often felt for pool cars, its TCO possibilities do not fit with the acquisition costs of EVs.

Hypothesis 1a TCO fit positively influences purchase intention of fleet managers.

The same holds true for functionality attributes of cars. Depending on their business model, companies face different kinds of mobility needs that need to be fulfilled. For example, a sales person needs to drive several hundred kilometers a day, whereas employees nursing people at their home often drive in an inner-city radius. Therefore, the business model mobility fit is influential on the decision to adopt EVs.

Hypothesis 1b Mobility-business-model fit positively influences purchase intention of fleet managers.

Cars may suit as marketing instrument for companies. Especially pool cars are often branded with company logos. Therefore, companies might see relevance in communicating a certain kind of image with their cars. For example, employees of luxury brands would communicate an incongruent brand appearance when driving to their customers in a small economy vehicle. Moreover, especially with environmentally friendly cars, companies try to enhancing their ecological footprint and like to demonstrate this to their customers.

Hypothesis 1c Symbolic fit positively influences purchase intention of fleet managers.

(Lieven et al. 2011) put driving range and acquisition costs on a continuum, where one aspects can be more important than the other. In line with their findings, we propose that any improvement of either price or range enhances purchase intention of fleet managers significantly, but still not on the same level as ICEVs. An elevation of the range and a decrease in price on the same level as ICEVs on the one hand improves purchase intention significantly in comparison to current EVs, and on the other hand, improves purchase intention to a similar level as ICEVs.

Hypothesis 2a An improvement of either driving range or acquisition price improves purchase intention of fleet managers significantly, compared to current EVs.

Hypothesis 2b An improvement of both driving range and acquisition price improves purchase intention of fleet managers significantly, compared to current EVs, so that purchase intention between ICEVs and improved EVs are not significantly different from one another.

In order to avoid variance attributed to different fleet types, we controlled for the number of cars and of pool cars, since it is assumed the larger fleet and the larger pool fleet there is, the more possibility the company has to substitute range-related drawbacks of EVs with ICEVs. Moreover, we controlled for financing, i.e. whether the company usually buys their cars or leases them. As we argued in our scenario mainly about the acquisition price, we avoided distortions due to different financing types by including this control variable. We checked for the mobility needs of firm by controlling for the usage frequency of pool cars, their range being driven on average on a daily basis and their frequency that their cars are driven for distances above 100 km. To avoid distortion based on their business model, we included industry as control variable. To check for the importance of environmental friendliness we included variables on environmental friendliness of the fleet, company and corporate CO_2-regulations related to fleet vehicles.

3.4.2 Operationalization

A scenario-based 2 × 2 between-subjects experimental design has been chosen, where acquisition price and driving range has been manipulated. The price was chosen to be either 18,000 or 28,000 €, whereas the range was either 190 or 600 km. This corresponds roughly with the actual prices and driving ranges with one tankful of a Volkswagen eGolf or Golf (petrol). For the price we also include information of a discount for fleets of 20%, which has already been included in the price above. As EVs are often seen as a way to improve carbon footprint, we did not include symbolic attributes in the manipulations, but rather held them constant, so that no bias can be related to this aspect.

We implemented a control scenario with a Golf run by petrol. The scenario description was for all respondents the same. They were told that they ought to buy a new compact class car for their pool fleet. Then they got offered the car, which has been manipulated in price and range. An exemplary scenario of the EV with both improved price and range can be seen in Fig. 3.1.

You are faced with the procurement of a new vehicle for the **pool fleet** of your company. You have received several offers from different manufacturers for **compact cars.** You asked the dealers to include the following elements into their car offer: light alloy wheels, multifunctional leather steering wheel, air-conditioning with 2-zones temperature control, portable navigation and infotainment system. You furthermore asked for a particularly environmentally friendly car. You have received the following offer:

Volkswagen e-Golf
e-Golf 85kW (115 hp) 1-Gear automatic

Energy consumption combined	12,7kWh/100km
CO2-emissions combined	0 g/km
Efficiency Class	A+
Price for the chosen equipment and motorization including optional extras	**€ 18.000,-**
Leasing rate (36 monthly rates à)	**€ 200,-**

*Drive now up to **600 km** locally emissionfree. You can easily charge your vehicle at any electrical outlet or at a wallbox available here. Not more expensive as a comparable Diesel – order now!*
Your discount for fleet customers is already taken into account.

Fig. 3.1 Example Scenario

Respondents were first asked about their task in the fleet purchasing process, followed by questions concerning the fleet (size, average daily driving range, etc.). Then the scenario was presented followed by questions regarding their purchase intention. At the end of the questionnaire, manipulation checks have been included. We used 7-point Likert scales for all latent variables, which we adopted from the literature (TCO and mobility fit: (Cable and deRue 2002); symbolic fit: (Lee et al. 2013); environmental friendliness of fleet and company: (Banerjee 2002; Banerjee et al. 2003); purchase intention: (Dodds et al. 1991)). All model and construct fit criteria were deemed to be acceptable. The correlation matrix can be seen in Table 3.1.

3.4.3 Sample Description

To answer the research questions an online based survey has been conducted. The link has been sent to roughly 3000 companies located in Germany during August 2015 and January 2016. Only 113 questionnaires returned, of which 7 had to be eliminated after quality checks, so that the total sample consists of 107 respondents. To participate in the survey, respondents had to be responsible in the fleet management of the firm. The survey lasted on average 8 min.

On average the respondents were 10.75 years employed in fleet management and took care of Ø 880 cars, of which Ø 66 cars belonged to the pool fleet. 66% of the pool cars were used daily for several times. Only 14% of the cars were used one to

Table 3.1 Correlation Matrix

	(1)	(2)	(3)	(4)	(5)	(6)	(7)	(8)
(1) Purchase intention	1							
(2) Environmental friendliness of company	.075	1						
(3) Probability of EV purchase in future	.481**	.156	1					
(4) Financing	.070	.131	-.037	1				
(5) Usage frequency	.151	.124	-.018	.098	1			
(6) TCO-Fit	.534**	.016	.391**	.151	.385**	1		
(7) Mobility-business-model-fit	.633**	.099	.383**	.038	.438**	.489**	1	
(8) Symbolic fit	.604**	−231*	.339**	.004	.105	.441**	.495**	1

Significance **p < .01; *p < .05

Table 3.2 Sample Composition

Industry affiliation		Area of fleet management		Number of employees	
Industry	30%	Administration & purchase	31%	<50	42%
Services	24%	Administration only	9%	51–200	11%
Energy	8%	Purchase only	23%	201–500	11%
Other	33%	All tasks	30%	>500	36%
		Other	7%		

two times a week or less. Most of the cars were used for distances below 100 km (66.4%). However, pool cars were driven distances above 100 km only up to 10 times a year in 71% of the cases. 2/3 of the respondents used leasing models. An overview of the sample composition can be found in Table 3.2.

3.4.4 Hypotheses Testing

To test whether the TCO, mobility needs and image of the cars shown in the scenario fits to the company, a regression analysis was conducted, controlling for the same variables as in the experiment as well as for the scenarios (compared to the EV scenario with both future price and range). The variance in purchase intention is explained by 61.0%. Symbolic fit ($\beta = .291$, $p < .001$), mobility needs fit ($\beta = .313$, $p < .001$), as well as TCO-fit ($\beta = .280$, $p < .05$) all significantly explain purchase intention. Therefore, Hypothesis 1 is supported by our data.

To test our hypothesis that an improvement in either price or range significantly enhances purchase intention, a variance analysis was conducted for the five scenarios. We included the following control variables in order to improve model fit: usage frequency, intention to purchase an EV (independent of the one in the scenario) in the future, financing and environmental friendliness of company.

The test of between-subject-effects shows an explained variance of 34.0%. As expected, the purchase intention of the ICEV scenario (control scenario) and the EV scenario with current price and range (EV as it is today) are significantly different from one another ($p < .05$). The ICEV scenario is not significantly different to all other scenarios, where either price or range or both have been manipulated, indicating that either one has to be improved to elevate purchase intention on a similar level to ICEVs.

The results are shown in Figs. 3.2 and 3.3. Even though these results indicate that either one of the manipulations elevates purchase intention to a similar level as the one for the ICEV, we wanted to check for interaction effects of price and range to check whether price is indeed more important in enhancing purchase intention.

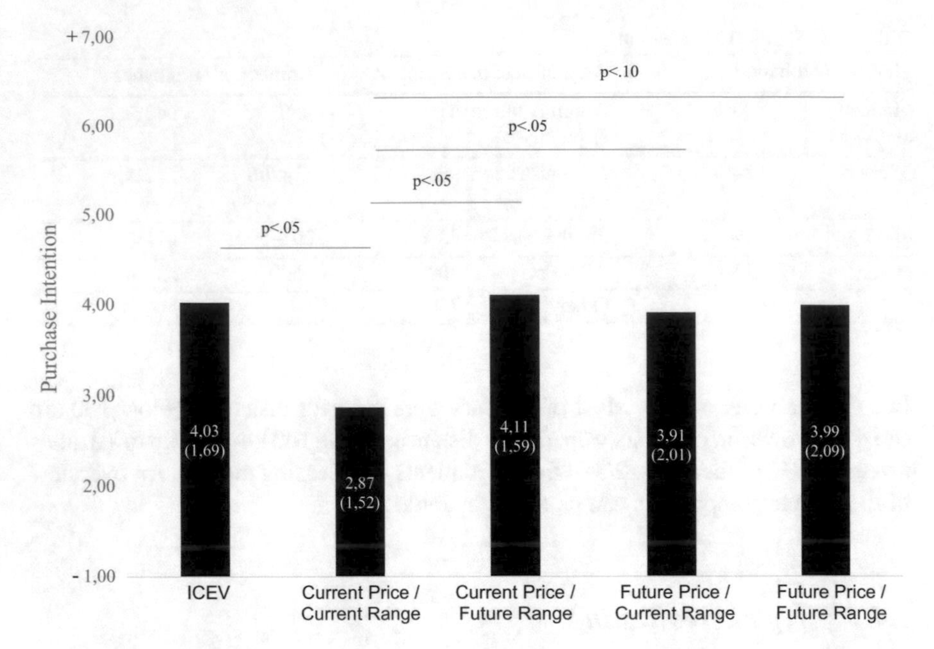

Fig. 3.2 Variance Testing

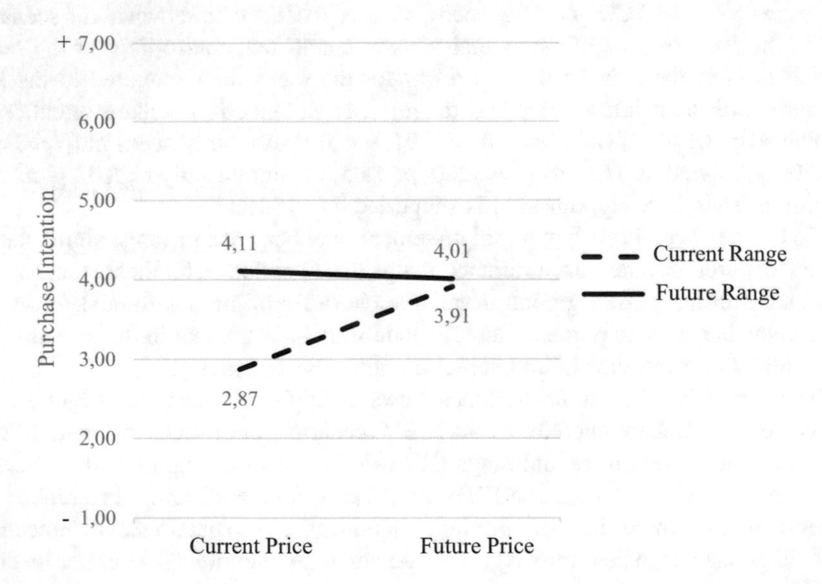

Fig. 3.3 Interaction Effect

The interaction of price and range is weakly significant (F = 3.700; p = .057). As a look at the means revealed a higher impact of range with the current price, we splitted the file to include only the two scenarios where range was manipulated. Here the difference in range was significant at a 1% level (F = 12.263).

We furthermore looked at the top-two boxes, i.e. all participants who had a purchase intention of 5.5 and above.[2] 26 fleet managers had a purchase intention of 5.5 and above, five of which were willing to buy the ICEV in the control scenario, which are 25% of the respondents seeing the control scenario. It became apparent that no participant wanted to buy the current EV (current price and current range). 28% of the people looking at the future EV-scenario, in which both price and range are at the same level as for ICEVs, had a high purchase intention. For the two other scenarios in which one of the manipulations was kept to the level of current EVs, a high purchase intention was expressed by 32% (current price and future range) and respectively 40% (future price and current range) of the respondents of the respective scenario.

3.5 Discussion and Implications

The findings of the preliminary study as well as the study of Szierchula (2014) could be validated. EVs need to fit on the TCO, mobility and symbolism dimension in order for fleet managers to have high purchase intention. The most interesting findings stem from the experiment where it has been shown that an improvement of either price or range affects the purchase intention significantly in a positive direction. In comparison of the car configuration of current EVs, it is enough to either increase range at the same acquisition costs or the price at the same driving range. This difference can be explained by the different business models of companies and their particular use of vehicle. For example, some companies operate on a smaller radius than others, such as urban logistics and services, whereas others drive hundreds of kilometers a day, such as sales people.

This study makes several contributions to the literature on fleet vehicle adoption in general and EV adoption in particular. It enhances the knowledge about relevant factors of organizational vehicle decision-making by validating the qualitative and descriptive findings of current research. Thereby, we broaden the scope of variables being used to determine EV adoption by using variables related to the fit of the vehicles besides the perceived usefulness and ease of use of studies applying the Technology Acceptance Model. As Sierzchula (2014) found, companies adopt EVs for firm-specific reasons, we found that the image of the car largely influences the car choice. So car manufacturers and dealers must be aware to carefully transmit a certain kind of image when marketing their EVs.

We found that the improvement of either range to the same level as ICEVs or the price to the same level as ICEVs has already significant impact on the purchase intentions of fleet managers, so that fleet managers appear to be indifferent into buying an EV or an ICEV. This result suggest that car manufacturers must improve

[2]Since the construct purchase intention is represented by the mean value of four different items, 5.5 (rounded up to 6) has been considered to represent the lower top-two-box.

the driving range of EVs significantly or reduce the price, depending on the particular requirement of different companies. Financial subsidies from the government would not increase purchase rates significantly, as reported by (Kieckhäfer 2013).

3.6 Limitations and Future Research

We designed a small experiment where we improved price and range on the same level as in the control group scenario. It is therefore unclear whether small improvements in range and price have a similar effect. Future research should therefore vary driving range and acquisition price to find the optimal point where purchase intention for EVs does not significantly differ from ICEVs. Knowledge of the optimal selling price and range would help car manufacturer to realistically design a car with improved features. Our findings suggest that customer requirements are heterogeneous. Therefore, future research must empirically test the levels of price and range combinations in their influence of purchase intention. This information gives implications for car manufacturers as to which problem to tackle first and to what extent.

We did not include charging infrastructure and duration in our analysis, neither as control variable nor as manipulation. However, the possibility to install a charging infrastructure and to assign specific parking spaces for EVs is not always given for each and every company. Therefore, future studies should at least control for the possibility of installing a charging infrastructure at their grounds. Besides, the charging duration may be a problem for some companies as well. Especially when cars also need to run at night, long recharging duration could be a major barrier for adoption.

References

Banerjee SB (2002) Corporate environmentalism. The construct and its measurement. J Bus Res 55:177–191

Banerjee SB, Iyer ES, Kashyap RK (2003) Corporate environmentalism: antecedents and influence of industry type. J Mark 67(2):106–122

Cable DM, deRue DS (2002) The convergent and discriminant validity of subjective fit perceptions. J Appl Psychol 87(5):875–884

Contestable M, Offer GJ, Slade R, Jaeger F, Thoennes M (2011) Battery electric vehicles, hydrogen fuel cells and biofuels. Which will be the winner? Energy Environ Sci 4:3754–3772

Dodds WB, Monroe KB, Grewal D (1991) The Effects of Price, brand, and store information on buyers' Product Evaluations. J Mark Res 28(August):307–319

Doll C, Gutmann M, Wietschel M (2011) Integration von Elektrofahrzeugen in Carsharing-Flotten. Fraunhofer Systemforschung Elektromobilität FSEM

German Federal Environmental Agency (2016) Treibhausgas-Emissionen in Deutschland. Retrieved from http://www.umweltbundesamt.de/daten/klimawandel/treibhausgas-emissionen-in-deutschland (24.08.2016)

Gnann T, Plötz P, Wietschel M (2015) What is the market potential of plug-in electric vehicles as commercial passenger cars? A case study from Germany. Published in Transp Res Part D 37:171–187

Grausam M, Parzinger G, Müller U (2015) Handlungsempfehlungen zur Intgeration von Elektromobilität in Flotten für Fuhrparkbetreiber. BMVI, Berlin

Hagman J, Ritzén S, Stier JJ, Susilo Y (2016) Total cost of ownership and its potential implications for battery electric vehicle diffusion. Res Transp Bus Manag 18:11–17

Hiermann G, Puchinger J, Ropke S, Hartl RF (2016) The electric fleet size and mix vehicle routing problem with time windows and recharching stations. Eur J Oper Res 252(3):995–1018

Hindrue MK, Parsons GR, Kempton W, Garnder MP (2011) Willingness to pay for electric vehicles and their attributes. Resour Energy Econ 33(3):686–705

Holtsmark B, Skonhoft A (2014) The Norwegian support and subsidy policy of electric cars. Should it be adopted by other countries? Environ Sci Policy 42:160–168

Kaplan S, Gruper J, Reinthaler M, Klauenberg J (2016) Intentions to introduce electric vehicles in the commercial sector: a model based on the theory of planned behaviour. Res Transp Econ 55:12–19

Kickhäfer K (2013) Marktsimulation zur strategischen Planung von Produktportfolios. Dargestellt am Beispiel innovativer Antriebe in der Automobilindustrie, Springer Gabler, Wiesbaden

Kley F, Lerch C, Dallinger D (2011) New business models for electric cars—A holistic approach. Energy Policy 39:3392–3403

Kraftfahrtbundesamt (KBA) (2016) Neuzulassungen von Pkw im Jahr 2015 nach privaten und gewerblichen Haltern. Retrieved from http://www.kba.de/DE/Statistik/Fahrzeuge/Neuzulassungen/n_jahresbilanz.html?nn=644522 (24.08.2016)

Lane B, Potter S (2007) The adoption of cleaner vehicles in the UK: exploring the consumer attitude-action gap. J Clean Prod 15:1085–1092

Lee EM, Park SY, Lee HJ (2013) Employee perception of CSR activities: Ist antecedents and consequences. J Bus Res 66:1716–1724

Lieven T, Mühlmeier S, Henkel S, Waller JF (2011) Who will buy electric cars? An empirical study in Germany. Transp Res Part D 16:236–243

Müller D, Ommen N, Woisetschläger DM (2015) Ein Segmentierungsansatz für die Adoption von Elektrofahrzeugen in Unternehmen. In: Proff H, Heinzel A, Leisten R, Schmidt A, Schönharting J, Schramm D, Witt G (Hrsg.): Entscheidungen beim Übergang in die Elektromobilität - Technische und betriebswirtschaftliche Aspekte. Springer Gabler, Wiesbaden

Nesbitt K, Sperling D (2001) Fleet purchase behavior: decision processes and implications for new vehicle technologies and fuels. Transp Res Part C 9:297–318

Showcase Electric Mobility Germany (2014) Schaufenster Elektromobilität. Retrieved 15.09.2016 at http://schaufenster-elektromobilitaet.org/de/content/ueber_das_programm/foerderung_schaufensterprogramm/foerderung_schaufensterprogramm_1.html

Sierzchula W, Bakker S, Maat K, van Wee B (2012) Technological diversity of emerging eco-innovations: a case study of the automobile industry. J Clean Prod 37:211–220

Sierzchula W (2014) Factors influencing fleet manager adoption of electric vehicles. Transp Res Part D 31:126–134

Wu G, Inderbitzin A, Bening C (2015) Total cost of ownership of electric vehicles compared to conventional vehicles: a probabilistic analysis and projection across market segments. Published in Energy Policy 80:196–214

Chapter 4
Planning of the Energy Supply of Electric Vehicles

Jan Mummel, Michael Kurrat, Ole Roesky, Jürgen Köhler and Lorenz Soleymani

4.1 Introduction

In the project Fleets Go Green innovative concepts have been developed to improve the integration of electric vehicle fleets into the electric power system and to increase the proportion of energy from renewable sources in their supply. The charging should be "green" (using sustainable energy), efficient and convenient for the users.

Coupling charging stations for electric vehicles with local energy production has a high potential to enhance the integration of renewable energy sources. This local supply is an additional expansion of renewable energy sources. Moreover, the local coupling to charging stations has the advantage that power losses are avoided, local distribution networks are relieved and the development of renewable energies is accelerated. This also increases the ecological potential of the fleets in contrast to the use of a conventional electricity mix.

An alternative to the use of additional local renewable energy sources is the use of electricity from existing plants. This energy is currently fed into the grid and consumed. However, being used primarily for charging electric vehicles, this power cannot be used for the supply of existing consumers anymore. Consequently, the new demand exceeds the existing supply. An additional expansion of capacity is therefore useful from an environmental perspective.

J. Mummel (✉) · M. Kurrat · L. Soleymani
Institute of High Voltage Technology and Electrical Power Systems - elenia, Technische Universität Braunschweig, Braunschweig, Germany
e-mail: j.mummel@tu-braunschweig.de

O. Roesky
TLK-Thermo GmbH, Braunschweig, Germany

J. Köhler
Thermal science laboratory (IfT), Technische Universität Braunschweig, Braunschweig, Germany

C. Herrmann et al. (eds.), *Fleets Go Green*, Sustainable Production, Life Cycle Engineering and Management,
https://doi.org/10.1007/978-3-319-72724-0_4

The infrastructure model introduced in this paper is used to determine and map the infrastructure of industrial fleets. It allows setting a charging strategy, the assessment of the electricity mix, the environmental impact and the cost of fleet operation during charging of electric vehicles. It supports the targeted integration of electric vehicles into the company fleet, taking into account the local generation capacity, power restrictions and economic and environmental objectives of a fleet company.

In addition to the strategic planning of the charging and energy infrastructure, the integration of electric vehicles requires an operational charge control. This, furthermore, facilitates dynamic electric prices. In the final part of this paper a concept of operational charging control for a central controlled charging management is presented.

4.2 Concepts for the Strategic Planning of the Energy Supply of Electric Vehicles Fleets with Local Production of Renewable Energy

4.2.1 Model Introduction

The aim of this model is to support the fleet or infrastructure manager in the strategic planning of the charging and local energy infrastructure.

For this purpose, potential energy sources are determined and the optimized size of the system considering the electric grid capacity limit and the contribution of the local energy sources is calculated.

The model is divided into three modules. In the first module, the relevant data are prepared for the calculation of the charging schedules. The parameters are the maximal power of the public grid connection, the local generation plants and the driving information.

Module 2 calculates, in consideration of the parameters set in module 1, charging schedules for the considered vehicle fleet. During this process, controlled and uncontrolled charging are distinguished. This is necessary for the selection of appropriate infrastructure.

The final module (module 3) evaluates the results obtained and derives a recommendation for the fleet operator. A flow chart of the model is shown in Fig. 4.1, followed by a detailed description of the three different modules.

4.2.2 Data Preparation

In module 1, the application scenario for the subsequent simulation is determined. For this purpose, three categories are taken into consideration, namely the vehicle fleet itself, the company operating the fleet as well as the energy generation plants

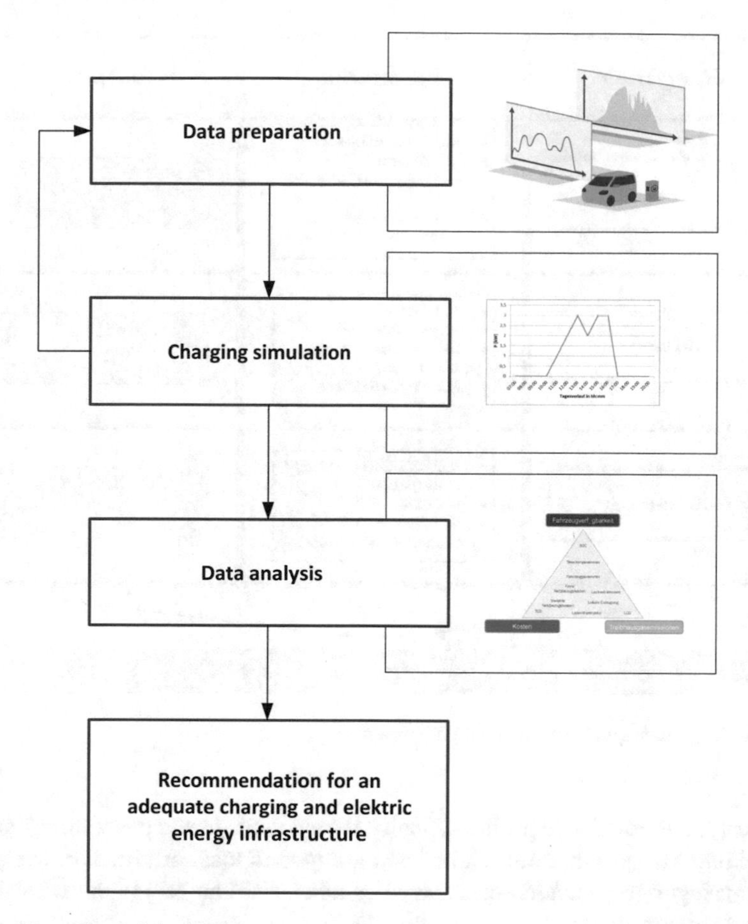

Fig. 4.1 Flow Chart of the Model

potentially used for the power supply. These categories can be freely combined with each other. Each category offers a set of different parameters extrapolated from synthetic and measured profiles. The configuration of the application scenario is performed within a specified tool chain through the stepwise definition of the required parameters. The program sequence of the data preparation is shown in Fig. 4.2.

In the first step, a vehicle fleet is set up. For this purpose, the parameters vehicle, driving profile, driving times and driving distances are entered into the system. When calculating the total energy consumption of the vehicles, secondary energy consumptions are taken into account depending on the time of the year (see section charging efficiency). Furthermore, the loss of the charging infrastructure is considered. To this end, several measurements of the energy losses for the application scenarios 2.3, 7.2 and 22 kW AC are taken (see section charging efficiency). In the second step, the most important characteristics of the company under consideration are determined. These include, among others, the company's sector, number of employees as well as

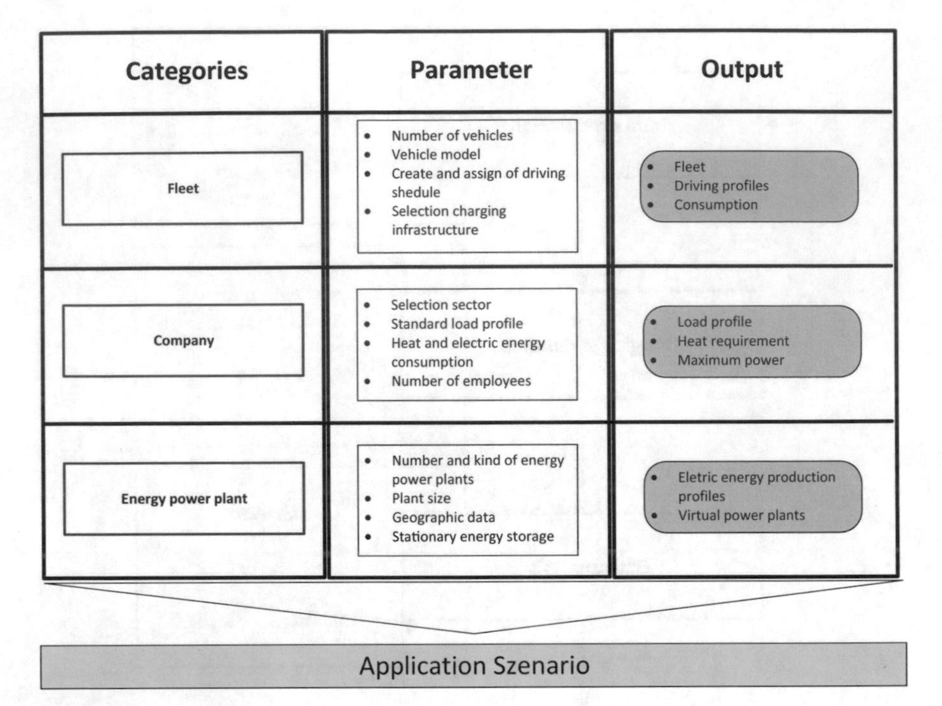

Fig. 4.2 Program sequence of the data preparation

electrical and thermal load profiles. Finally, the available power generation plants are entered into the system, dimensioned and grouped in local or virtual power plants. For this category parameters such as type, number, size and geographical location of the plant are considered.

With these input variables load and generation forecasts are generated for the simulation period. By combining these data sets a scenario is created, which will finally be simulated for the entire period.

4.2.3 Simulation

The simulation of the charging operations is implemented in MATLAB® and coupled to a database for the data exchange. The simulation calculates charging schedules for electric vehicles fleets. The linear optimization problem is applied to optimize the loading operations by minimizing the costs over the entire observation period. An essential step in the development for the solution of an optimization problem is the transfer of the real system in a formal model. In the following section, the established project approach is explained and the system model is outlined.

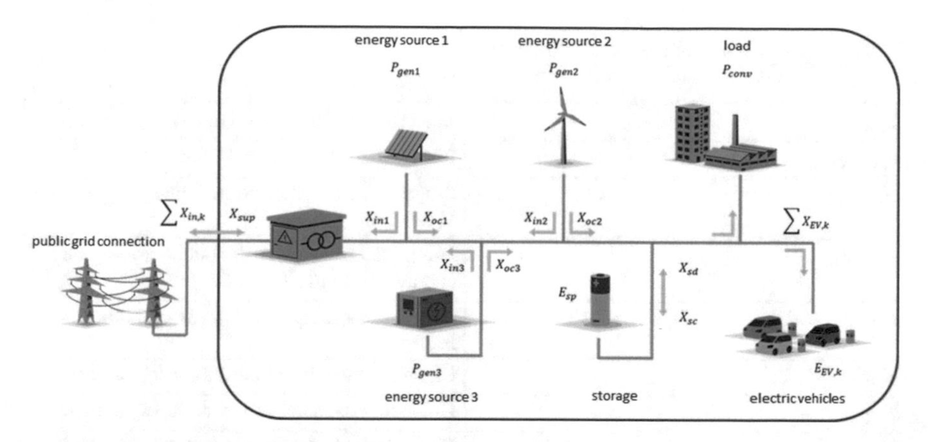

Fig. 4.3 Graphical figure to describe the optimization problem

Creating the Model

In the first step, the descriptive electric grid model underlying the optimization is generated. It is developed from a model published by Plota and Rehtanz (2014) which describes a local electric grid. The model consists of the load from the industrial company and the power generation plant, both connected via network connection point to the public electric grid. The production site is in a location of up to three production plants. The location is assigned to an electric vehicle fleet. Additionally, stationary energy storages can be integrated. On the public grid connection point energy can be fed in or excess energy can be obtained.

The power respectively energy flows with their corresponding directions are represented by green arrows in the model (Fig. 4.3) and are the essential parameters of the model. Obviously, the consideration of power and energy is equivalent in the optimization problem. Due to the time discretization of the optimization period both variables are linked via the length of the selected time interval Δt. For the time interval Δt the following equation is applied with E_t representing the energy and P_t the power during the time interval Δt:

$$\sum_{t=1}^{T} E_t = P_t * \Delta t$$

In Fig. 4.3, the parameters of the model are classified into three different types. The variable "P" indicates a power that is predefined in the module 2 for every energy plant and conventional load and, thus, is not part of the optimization variables. To state an example, this could be the infeed power of a photovoltaic system, which depends on the weather conditions (e.g. cloudiness). Electric power flows, which are denoted by "X", are part of the optimization variables. They are determined by solving the optimization problem. An example for this type of parameter is the charging power of an electric vehicle. The parameters described by the variable "E"

Table 4.1 Description of the model parameter

Name	Description
P_{conv}	Load (for example, through a company)
$P_{gen1,2,3}$	Power of energy source 1, 2 or 3
$X_{oc1,2,3}$	Proportion of consumption from energy source 1, 2 oder 3
$X_{in1,2,3}$	Part of electric feed in the public grid of the energy sources 1, 2 oder 3
$X_{EV,k}$	Charging capacity of k-ten vehicle from the fleet
X_{sc}	Charging capacity of the stationary storage
X_{sd}	Unloading capacity of the stationary storage
X_{sup}	Electric power demand from the public grid
$E_{EV,k}$	State of charge of the k-ten vehicle from the fleet
E_{sp}	State of charge of the stationary storage

are also part of the minimal cost optimization variables. They represent the energy content of a memory at the time under consideration. A description of the individual parameters is shown in Table 4.1.

Controlled Charging

The charge algorithm consists of several functions that fulfil different tasks in the program sequence. Figure 4.4 shows the program sequence. At first the user selects an application scenario, created in module 1. All associated data is collected from the database. Key information is displayed to the user. Subsequently, the data processing is carried out. Before running the optimization, the overall problem is divided into smaller sub-problems. The developed model is converted into a linear program. For the generated sub problems an objective function and a system of constraints are established. The problem is then solved using a linear optimization algorithm (Mummel et al. 2016).

Uncontrolled Charging

In the uncontrolled-charging part of the program there is no intervention from the charging management system. The charging function is therefore influenced by user behavior, the vehicle model and the charging infrastructure only. The charging process starts after a vehicle has been connected to the charging station. The process is terminated once the vehicle is fully charged, or upon manual interruption by the user. These calculations are performed after completion of the optimization algorithm, using identical input data.

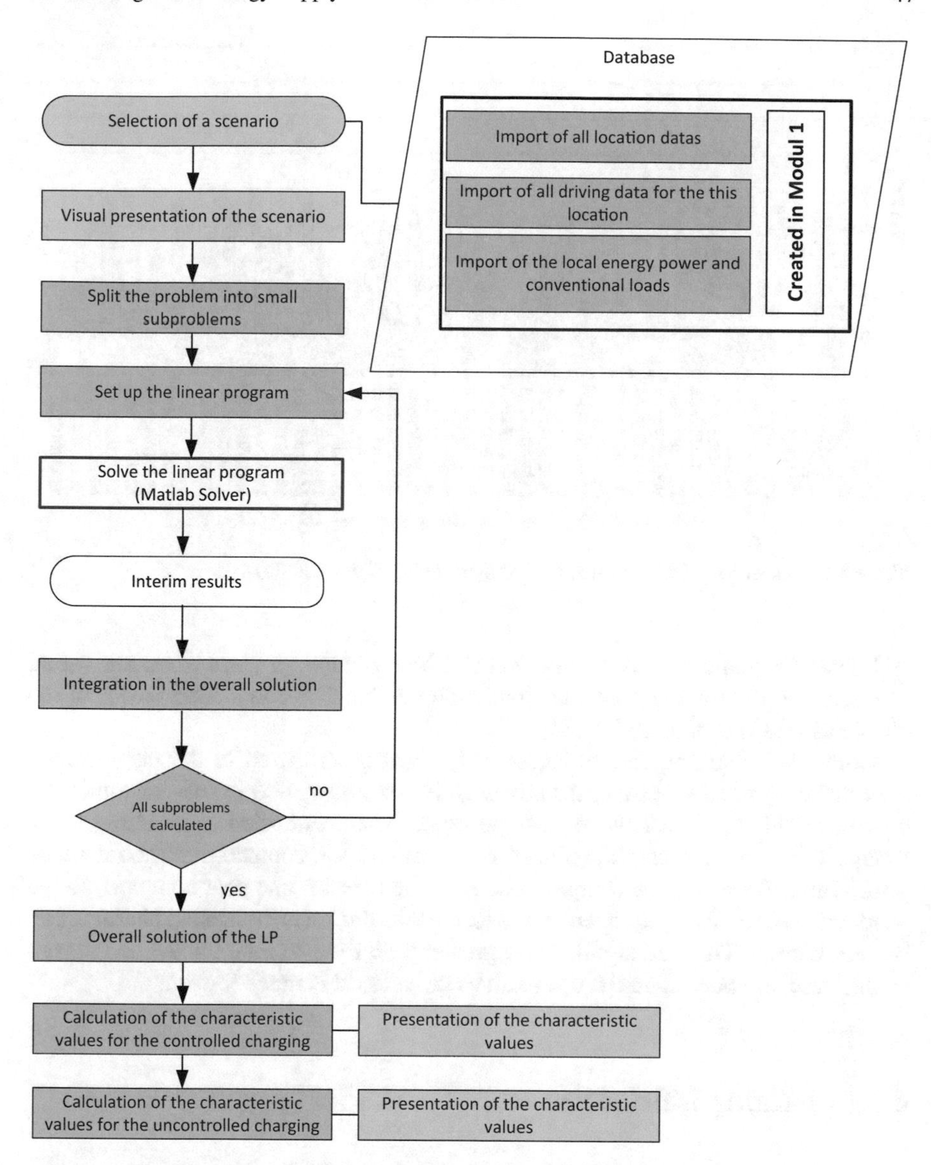

Fig. 4.4 Program sequence of the controlled charging

4.2.4 Application Scenario Example

The following application is used: a fleet with real driving profiles from the project Fleets Go Green and a PV system (10 kWp) as a power generator.

A total of five vehicles is considered. Three vehicles will be charged with up to 22 kW AC. The charging power for the other two vehicles is limited to 3.7 kW AC.

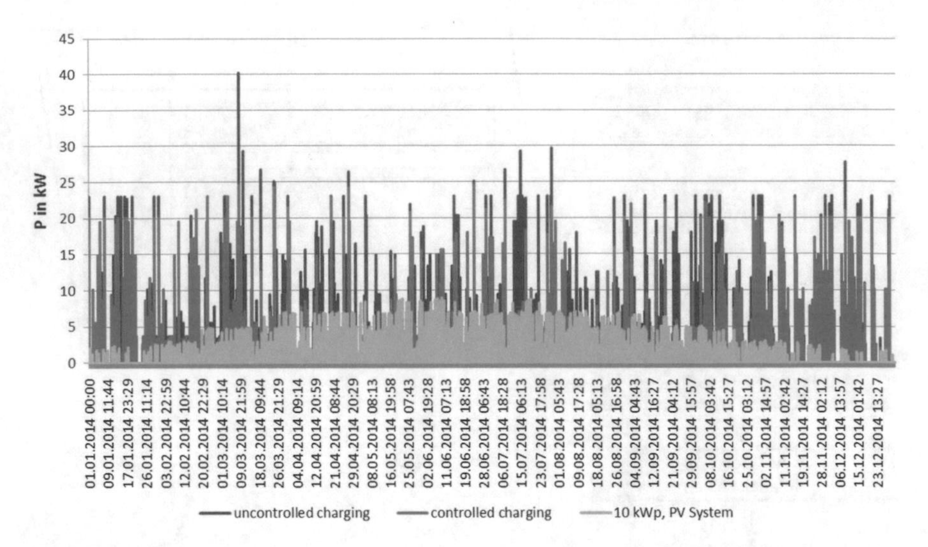

Fig. 4.5 Uncontrolled and controlled charging processes of the year 2014

AC charging points are used permanent. Power generation parameters are varied to determine the influence on the simulation results. The connection power of the electrical grid is limited to 20 kW.

Controlled charging causes larger CO_2 savings compared to the uncontrolled alternative. In the PV system, the coverage is just under 49% in case of controlled charging. This is almost twice as large as in the uncontrolled case. Using a PV system CO_2 emission can be reduced by about 850 kg, compared to the reference grid. Figure 4.5 shows the charging power of the fleet for one year compared to the production of the PV system. Here, the load peaks during uncontrolled charging can be seen clearly. They are significantly greater than the maximum generated power. In this case, the burden on the upstream electrical grid is much higher.

4.3 Charging Efficiencies

Within the research project, Fleets Go Green, about 30 electric vehicles have been obtained for the University of Braunschweig and the local energy provider BS|Energy. These vehicles are used by BS|Energy in a fleet operation and at the University in a pool-system.

The vehicles are equipped with high-voltage measuring equipment in order to trace and record transient data when driving or charging the vehicles. This data provides a detailed analysis of the mobility and charging behavior of users in fleet operation.

In contrast to conventional vehicles where all of the gasoline filled into the tank can be used for propulsion, refilling the batteries of an electric vehicle always involves

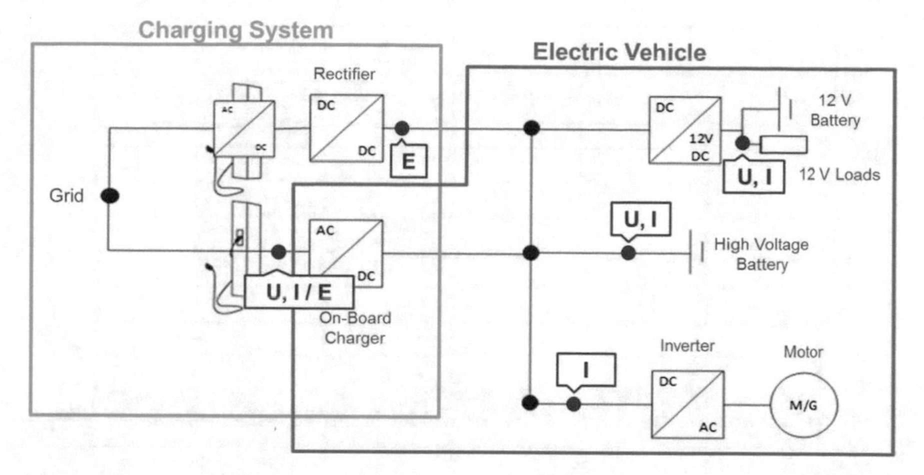

Fig. 4.6 On-board power supply of the vehicle and interface to the energy grid

losses, especially when using an AC charging station (Tober and Lenz 2016). On the one hand, the user of an electric vehicle has to pay for the energy lost in the transfer from the energy grid into the battery. On the other hand, the regulatory agencies have considered these losses when determining the overall efficiency of electric vehicles (ECE 2013). As the charging efficiency of a vehicle affects its overall efficiency, it should be of high priority for vehicle manufacturers.

4.3.1 Charging System and Electric Vehicle

The powertrain is the central element of an electric vehicle. It consists of a high voltage battery, the power electronics and an electric motor. The on-board power supply interconnects all electric components of the vehicle. Figure 4.6 shows a layout of the on-board power supply with infrastructure-connections as well as the places where measurements are taken.

The left side of the figure describes the charging system. For AC-charging (alternating current) the electric vehicle requires a one-, or three-phase inverter. For DC-charging (direct current) on the other hand, the charging system is entirely integrated in the charging station outside the vehicle.

The on-board power supply has a high voltage (HV) and a low voltage (LV) system. The power train and the charging system are high voltage components. Low voltage components are the on-board electric system consisting of control units, electrical pumps, fans, etc. The DC/DC-converter connects the vehicles high and low voltage systems and supplies energy to the low voltage side. During the charging process the losses of the on-board charger result in heat which has to be dissipated from the charger through a cooling circuit. The electrical pumps to run the cooling circuit and,

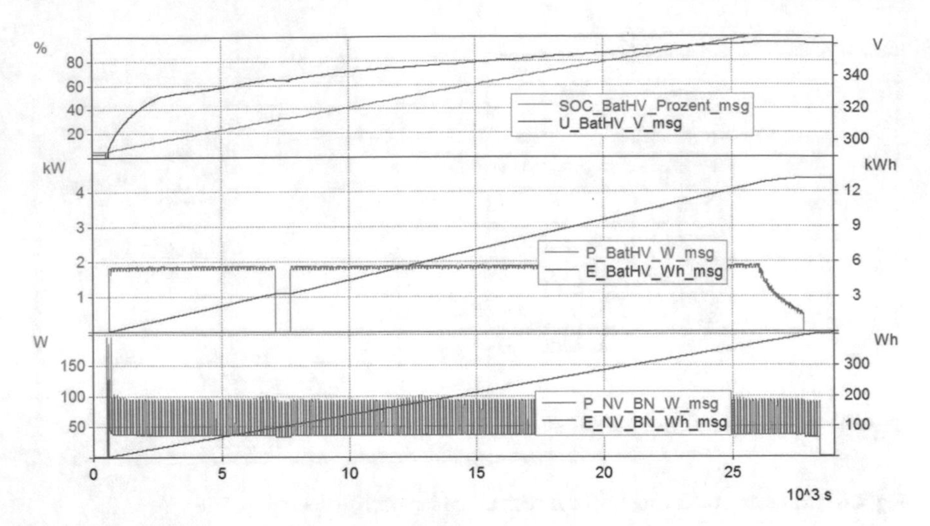

Fig. 4.7 Citroën C-Zero 2 kW charging at wall socket, transient data

in case of high outside temperatures, the radiator fan has to be supplied with energy. The high voltage components used while charging, have their own control units that also consume energy provided by the low voltage system (Kampker 2013).

4.3.2 Charging Measurements

Different charging solutions result in different costs for the charging infrastructure at home or in public spaces as well as in the costs for the on-board charging system. Moreover, charging at different power levels also results in different losses due to different charging efficiencies as well as charging times. Different charging measurements for different electric vehicles at different charging power levels have been conducted in order to quantify the advantages and disadvantages of different charging systems. Selected Measurements for a Smart ED and a Citroën C-Zero will be further detailed in this section. The vehicles differ in the technology used for charging. The Smart ED uses an on-board charger that allows AC charging with up to 22 kW. The Citroën C-Zero allows AC charging up to 3.6 kW and DC charging with up to 50 kW. Further specifications can be found in Table 2.1, Chap. 2. Figure 4.7 shows the transient data logged inside a Citroën C-Zero when charging at an ordinary wall socket at 2.0 kW (AC).

The onboard charging system inverts the alternating current (AC) from the grid to direct current (DC) to be used inside the vehicle´s high voltage DC system. The high voltage battery is then charged with 1.7 kW. Towards the end of the charging process, after around seven hours, the charging power is gradually being derated from 1.7 kW to around 0.5 kW and then terminated to protect the battery. While the current

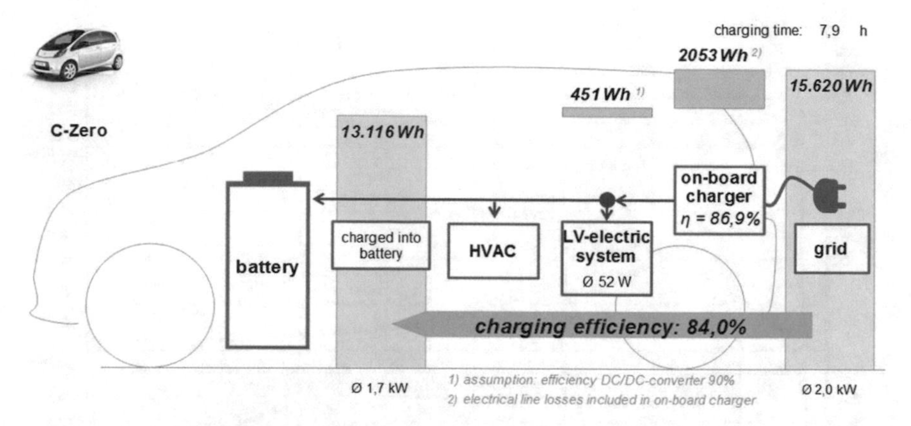

Fig. 4.8 Citroën C-Zero 2 kW charging at wall socket, cumulative data

charged into the battery stays relatively constant throughout the process, the battery voltage rises continuously. When the end-of-charge voltage is reached, the charging current is derated in order to not further raise the voltage in the battery cells. At the end 13,116 Wh have been charged into the battery over a period of approximately eight hours. Considering the 15,620 Wh provided by the grid, as shown in Fig. 4.8, this results in an overall charging efficiency of 84.0%.

The low voltage electric system consumes on average 52 W throughout the charging process. With the assumption of a 90% efficiency (März 2004) of the DC/DC converter, this results in an energy consumption of 412 Wh. This leaves losses of 2053 Wh to be assigned to the on-board charger as well as line losses to and from the charger. The diagram also shows a block for heating, ventilation, and air-conditioning (HVAC), which is not considered in this paper because no pre-conditioning of the passenger compartment has been conducted during these measurements. Figures 4.9 and 4.10 respectively display a charging process of the same vehicle when charging at a DC-quick-charging station.

After approximately 5 min the constant charging voltage is reached and the charging power is gradually derated from around 40–10 kW after around 25 min. The charging process is automatically terminated at a state of charge (SoC) of 80%. In order to obtain a fully charged battery the charging process has to be reinitiated twice. After one hour effective charging 12,305 Wh have been charged into the battery which results in a charging efficiency of 95.9% taking into account the 12,830 Wh provided by the DC charging station as shown in Fig. 4.10. The low voltage power requirement is on average 253 W and the line losses amount to 238 Wh.

As a comparison to the C-Zero DC quick charge measurement a measurement of a Smart ED capable of being charged at 22 kW by means of a 3-phase AC charging station shall be discussed. Figure 4.11 shows that before being able to take the maximum power of 20 kW into the battery the vehicle is first charged at a lower power level for a couple of minutes. After 20 min the vehicle is not able to process

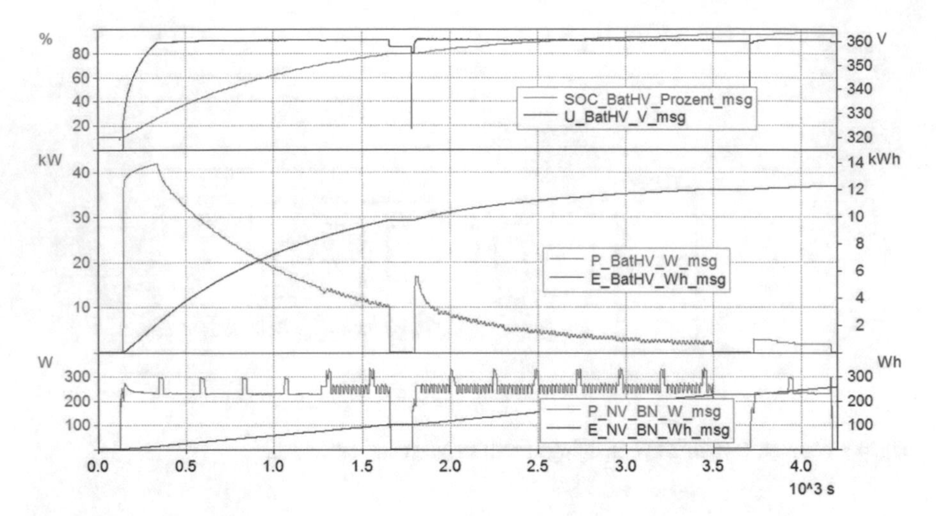

Fig. 4.9 Citroën C-Zero 50 kW charging at DC-station, transient data

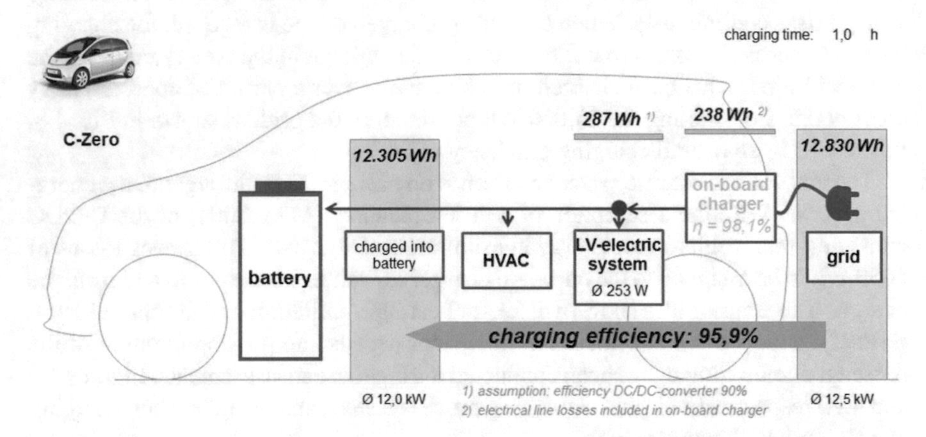

Fig. 4.10 Citroën C-Zero 50 kW charging at DC-station, cumulative data

the maximum power and after another 15 min the constant charge voltage is reached and the charging power is gradually decreased.

After one hour the charging process is terminated with 11,966 Wh charged into the battery. Figure 4.12 shows the energy distribution when charging at 22 kW AC. With 13,070 Wh provided by the grid this results in an overall charging efficiency of 91.6%. The low voltage electric system consumes on average 76 W throughout the charging process. This leaves losses of 1022 Wh to be assigned to the on-board charger as well as conduction losses to and from the charger.

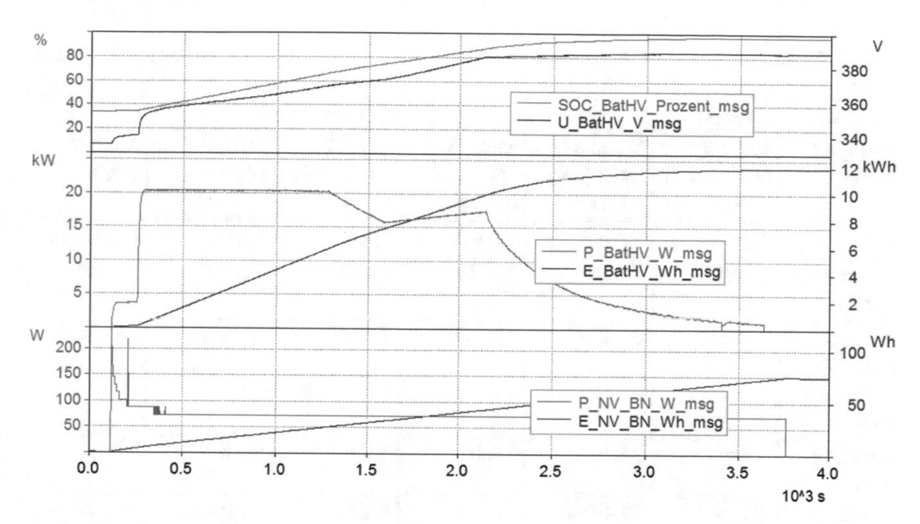

Fig. 4.11 Smart ED 22 kW charging at AC-station, transient data

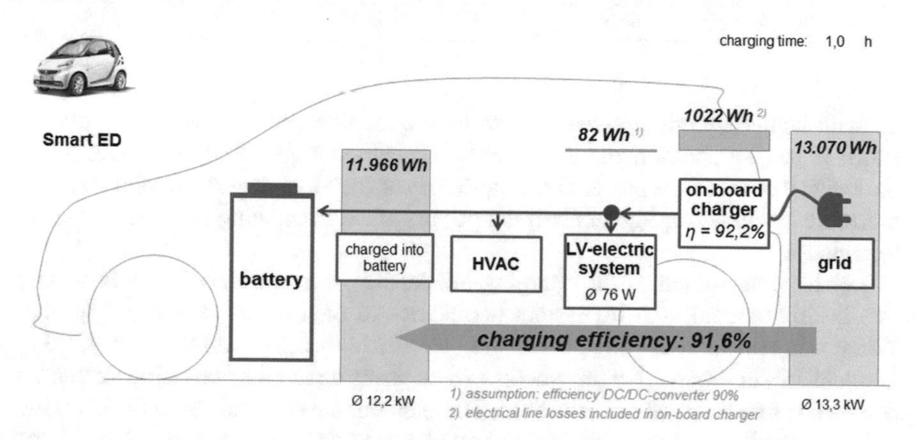

Fig. 4.12 Smart ED 22 kW charging at AC-station, cumulative data

4.3.3 Conclusion and Outlook

Table 4.2 shows the summarized comparison of charging measurements at different power levels. The measurements of the Smart ED not discussed in detail in this document can be found in Roesky et al. (2015) and Roesky (2016). For the considered vehicles it can be concluded that the overall charging efficiency rises with higher charging powers.

Charging the Citroën C-Zero at an ordinary wall socket with Type 2 yields the lowest charging efficiency of 84% whereas charging the same vehicle at a DC charging station through a CHAdeMO connection results in the highest charging efficiency of 95.9%. Furthermore, the power consumption of the low voltage electric system goes

Table 4.2 Summarized comparison of charging measurements at different power levels (Roesky 2016)

Vehicle	Smart ED			Citroën C-Zero		
Charging mode	Mode 2 (AC) (wall socket)	Mode 3 (AC) (only 10 A)	Mode 3 (AC) (not limited)	Mode 2 (AC) (wall socket)	Mode 3 (AC)	Mode 4 (DC)
Type	Type 2	Type 2	Type 2	Type 1	Type 1	CHAdeMO
Power level (Grid)	2.4 kW	7.2 kW	22 kW	2.4 kW	3.6 kW	50 kW
Average charging power in battery	2.3 kW	5.2 kW	12.2 kW	1.7 kW	2.4 kW	12.0 kW
Charging time	5.6 h	2.2 h	1.0 h	7.9 h	5.2 h	1.0 h
Charging efficiency	84.1%	90.4%	91.6%	84.0%	86.2%	95.9%
LV system	66 W	68 W	76 W	52 W	56 W	253 W

up with higher charging powers due to higher requirements. However, the overall shorter charging duration and the better efficiencies at higher charging power levels outweigh the higher LV power consumption. From an efficiency standpoint it should therefore be aimed for higher charging powers when connecting an electric vehicle to recharge.

For the comparison of the charging efficiencies, the considered system is only the electric vehicle with its system boundaries to the grid as shown in Fig. 4.6. When using an AC charging station the conversion from AC to DC takes place in the vehicle's on-board charger whereas when charging at a DC charging station no conversion inside the vehicle is needed. This explains the relatively high DC charging efficiencies when only considering the boundaries ex charging station to high voltage battery. Further research is needed to further investigate the real energy taken from the grid when charging at a DC station including the losses that occur inside the DC charging station itself. This evaluation would allow a more viable comparison of AC versus DC charging.

It can also be concluded that, considering vehicles with a battery capacity around 16–18 kWh, the overall charging time for a full charge does not differ significantly between a Mode 3 (AC) charge at 22 kW and a Mode 4 (DC) charge at 50 kW. This is due to the early derating of the charging power in the C-Zero to protect the battery from possibly damaging high currents towards the end of the charging process. With further development and vehicles with larger batteries the 50 kW DC quick charge stations will eventually yield shorter charging times.

The measurements have only been conducted with conductive charging technology. However, in the near future inductive charging will become technologically and

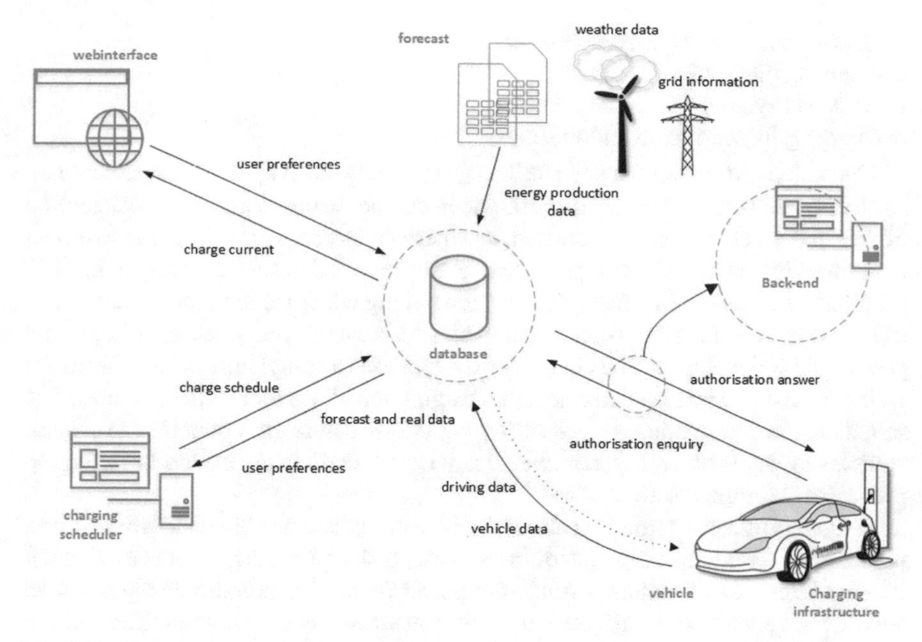

Fig. 4.13 Modular construction of *elias* Mummel et al. (2014a)

economically feasible so that additional losses entailed with this technology will also have to be taken into account.

4.4 Charging Management System

4.4.1 Introduction of System

elias is a modular based charging management system. It consists of several modules which use certain functions of charge planning. The system allows vehicle user to retrieve historical charging data, get actual information like available charging infrastructures and if necessary add intended drives. Fleet operators are able to add and edit several fleets, get historical and actual information about their fleets and if necessary do the drive scheduling for the next couple days. The configuration of the charging infrastructures can be done by the corresponding infrastructure operators. This includes the selection of the charging algorithm and maximum charging capacity in view of power grid restrictions.

The structure of the system is shown in Fig. 4.13. The central element of *elias* is a database which collects, stores and manipulates data from different subsystems. Thus, it serves as an interface between all other modules of the system which include:

- Web interface

- Grid-, weather-and generation forecast
- Charging plan tool
- Back end system
- Charging infrastructure and vehicles.

The web interface lies between the charging management system and users as well as fleet operators. Settings and preferences can be changed and drive scheduling of fleets and vehicles can be created. Furthermore, necessary information to plan charging of electric vehicles is presented. Based on weather forecast another module prognosticates renewable energy generation. Along with location specific data, such as load profile or maximum connected load, a forecast of energy balance of the local power grid is developed. The charging plan tool uses a compilation of data from the web interface and forecasts to calculate charging schedules for electric vehicles. The schedules are communicated to the charging infrastructure and connected to electric vehicles by the back end system. Feedback of vehicles occurs during the charging processes (Mummel et al. 2014a).

If there is either information about the type of vehicle or scheduled travel times, no charging schedule is calculated. In this case, a charge management is carried out without reference to mobility. Available power (sum of maximum connected load and power generation) is distributed to all connected electric vehicles according to various algorithms.

Due to the modular construction *elias* is scalable in number of locations, vehicles and in choice of the observation period. Different locations and fleets can be processed with a single system due the modularity. Therefore, *elias* can be seen as an aggregator of charging management systems. The structure of *elias* allows replacement of individual modules. For example, the forecasts of the electrical grid, weather or power generation can be replaced with external data or other strategies to generate charge schedules.

4.4.2 Process of Charging Management

The process of charging management in *elias* takes place in a fixed structure. Figure 4.14 shows a flow-chart of the charging management process by *elias*. The process is divided in three sequences. The sequence of data provision (blue) prepares input data for the charging schedule calculation by following programs. From an external data source, a weather forecast for location is obtained and stored on the data base. Along with other location based information a forecast for expected local energy production is calculated. There are specific characteristics for each generating plant used in this step. Furthermore, a load forecasting takes place which considers conventional consumer of the location.

The web interface is necessary to create, edit and visualise fleet and location specifications. In reality charge management users or superior fleet manager execute fleet planning for the following day. Travel time, route as well as vehicle and user

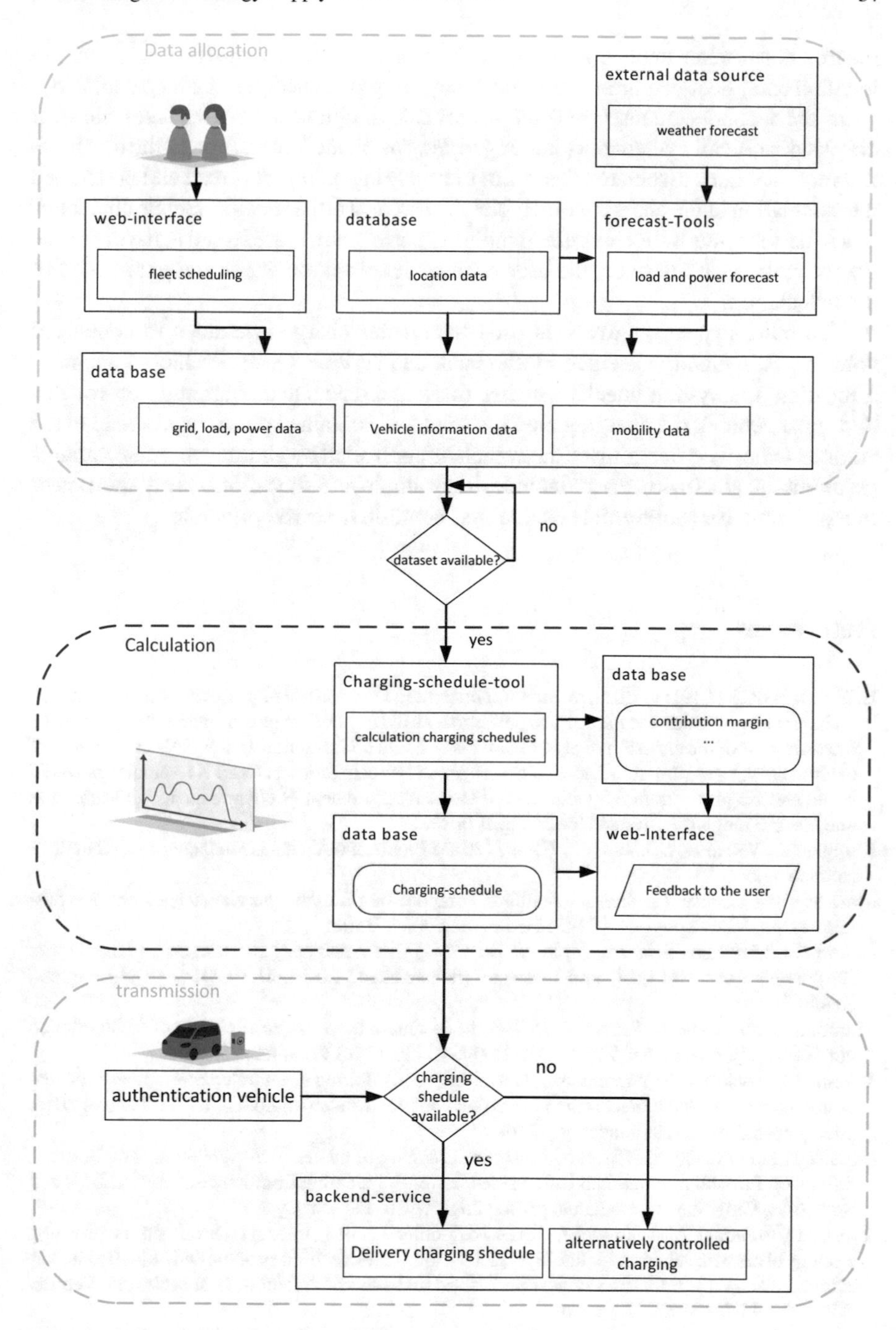

Fig. 4.14 Flow-chart of the charging management process

information is taken under consideration for each drive. Along with previously stored location data, fleet planning and forecasting charging schedules can be calculated.

In the second sequence (red) the actual calculation of schedules takes place. It checks periodically whether a planning order was created and stored in the database. If a request exists, associated data from the charging plan tool is read and performed the calculation of the charging timetables. These, as well as specific vehicle characteristics like the contribution margin or mobility performance, are stored in the database. By using the web interface, the user receives feedback on status and parameters of the calculation.

The third sequence (green) is used to transfer charge schedules to connected vehicles. It is mainly executed by the back end service. Once a vehicle logs on to a location, the system checks whether there is a calculated charging plan for this particular vehicle. If a charging plan is stored, the contained power is provided to the vehicle. Otherwise charging of the vehicle is controlled by alternative, static variants (Mummel et. al. 2014b). These include, for example, the allocation of available power in equal parts for each vehicle or in a first-come-first-served principle.

References

ECE (2013) ECE R-101: uniform provisions concerning the approval of passenger cars powered by an internal combustion engine only, or powered by a hybrid electric power train with regard to the measurement of the emission of carbon dioxide and fuel consumption and/or the measurement of electric energy consumption and electric range, and of categories M1 and N1 vehicles powered by an electric power train only with regard to the measurement of electric energy consumption and electric range E/ECE/324/Rev.2/Add.100/Rev.3

Kampker A, Vallée D, Schnettler A (Hrsg) (2013) Elektromobilität - Grundlagen einer Zukunftstechnologie

März M (2004) Leistungswandler – Schlüsselkomponenten für das Energiemanagement in Kraftfahrzeugen, VDE Kongress - GMM Fachtagung 2004, Berlin

Mummel J, Diekmann S, Kurrat M, Engel B (2014a) IKT-Anbindung für gesteuertes Laden unter Berücksichtigung von Last- und Erzeugungskapazitäten., 20.10.-21.10.2014, VDE-Kongress, Frankfurt

Mummel J, Soleymani L, Kurrat M (2014b) Ladesteuerungskonzepte für Elektrofahrzeugflotten zur Weiterentwicklung der Verteilnetze 18.09.-19.09., NEIS Konferenz, Hamburg

Mummel J, Stocklossa T, Wijtenburg J, Kurrat M (2016) Beitrag dezentraler Erzeugungseinheiten zum nachhaltigen und wirtschaftlichen Betrieb von Elektrofahrzeugflotten, 10.-12. Februar 2016, 14. Symposium Energieinnovation, Graz

Plota E, Rehtanz C (2014) Impact of controlled charging of an electrical vehicle fleet on business efficiency. In: 49th International Universities Power Engineering Conference (UPEC), 2014: 2–5 Sept 2014, Cluj-Napoca, Romania; proceedings. IEEE, Piscataway, NJ

Roesky O, Bodmann M, Mummel J, Kurrat M, Köhler J (2015) Impact of losses on the charging strategy of electric vehicles. In: Intelligente Transport- und Verkehrssysteme und –dienste Niedersachsen e.V. (Ed.). In: 12th Symposium Hybrid and Electric Vehicles, Braunschweig, Februar 2015, ISBN 978-3-937655-35-2.b

Roesky O (2016) Entwicklung einer Thermomanagement- und Verbrauchsoptimierten Ladestrategie für elektrifizierte Fahrzeuge. In: Großmann, H. (Hrsg.): 5. VDI-Fachkonferenz Thermomanagement für elektromotorisch angetriebene PKW 15. und 16. November 2016. Stuttgart : VDI Wissensforum

Tober W, Lenz, HP (Hrsg.) (2016) Praxisbericht Elektromobilität und Verbrennungsmotor - Analyse elektrifizierter Pkw-Antriebskonzepte. Wiesbaden : Springer Vieweg, 2016 — ISBN 978-3-658-13601-7

Chapter 5
Life Cycle Assessment of Electric Vehicles in Fleet Applications

Antal Dér, Selin Erkisi-Arici, Marek Stachura, Felipe Cerdas, Stefan Böhme and Christoph Herrmann

5.1 Introduction

In the view of the threat of climate change, environmental issues represent a raising concern for policy makers and society. While the transportation sector was responsible for 24% of worldwide anthropogenic greenhouse gas emissions in 2015, road transport accounted for 75% of total transport related emissions (IEA 2017). Despite efforts to limit the emissions from the transportation sector, global emissions grew by 68% between 1990 and 2015. Hence, vehicle manufacturers have to develop and produce vehicles in a more strongly regulated market environment. In the last decades, automotive companies focused on improving the drivetrain efficiency, developing alternative fuels and alternative drivetrain concepts like the electrification of vehicles due to stricter regulations and to a rising awareness of a more sustainable mobility within society. The electrification of the powertrain has the potential to decrease direct tailpipe emissions from vehicles. However, individual mobility with electric vehicles depends on other systems, e.g. on the energy supply and charging infrastructure. Thus, manifold interrelationships arise and different factors of influence need to be considered when assessing the environmental performance of electric vehicles.

A. Dér · S. Erkisi-Arici · F. Cerdas · S. Böhme (✉) · C. Herrmann
Institute of Machine Tools and Production Technology (IWF), Technische Universität Braunschweig, Braunschweig, Germany
e-mail: stefan.boehme@tu-braunschweig.de

M. Stachura
IPoint-Systems GmbH, Vienna, Austria

C. Herrmann et al. (eds.), *Fleets Go Green*, Sustainable Production, Life Cycle Engineering and Management,
https://doi.org/10.1007/978-3-319-72724-0_5

Around 66% of annual new passenger car registrations in 2015 in Germany are accounted for by companies and self-employed people (KBA 2015). Moreover, corporate vehicles have a higher annual mileage compared to privately used vehicles (Kasten et al. 2011). This qualifies corporate vehicle fleets for inducing a rapid diffusion of electric vehicles into the market and reducing road transport related CO_2 emissions. While electric vehicles have zero tailpipe CO_2 emissions during the use stage, their environmental burdens may be underestimated if other influencing factors and life cycle stages are neglected. For instance, the production of traction batteries poses big challenges from an environmental perspective due to its weight and required metals, whose production is linked to significant upstream activities. Consequently, the production of an electric vehicle is estimated to cause up to two times more environmental impact than the production of a conventional vehicle with an internal combustion engine in a given vehicle category (Hawkins et al. 2013; Cerdas et al. 2018). Besides, influencing factors related to vehicle production, different factors in the use stage can have a significant contribution to the environmental impact of electric vehicles that are beyond of today's scope of vehicle manufacturers. Regional factors such as climate, topography and electricity mix for charging have a significant effect on the environmental impact of electric vehicles (Egede et al. 2015; Li et al. 2016).

In order to achieve a more consistent understanding of the potential environmental impacts of electric vehicles, the afore-mentioned interdependencies and influencing factors need to be properly considered. Within the context of the project Fleets Go Green, this chapter presents the results of a Life Cycle Assessment done to evaluate the potential environmental impacts of electric vehicles in fleet operations. Life cycle inventories based on vehicle manufacturer's data and real world field test driving data have been collected. Furthermore, driving patterns and the interaction of ambient temperatures are considered, which provide insights of the influence of driver and region related factors on the energy demand of electric vehicles. The results help to better understand the environmental impacts of electric vehicles in fleet operations and support the derivation of recommendations.

5.2 Life Cycle Assessment in the Context of Electric Vehicles

5.2.1 Life Cycle Assessment Methodology

Life Cycle Assessment (LCA) is a standardized methodology for assessing and quantifying the environmental impacts associated with the use of a product or service over its entire life cycle (ISO 14040:2006 2006). The first step of the LCA is the Goal and Scope Definition. This step focuses among others on clarifying the decision context and the targeted audience, on defining system boundaries and on choosing a functional unit that represents the baseline for comparisons. Furthermore, relevant impact categories and impact assessment methods need to be chosen (Bjørn et al. 2018a, c).

The following step is the Life Cycle Inventory (LCI) Analysis that encompasses the collection of data and modelling of the energy, resource waste and emission flows within the product system and in exchange with the surrounding techno- and biosphere (Bjørn et al. 2018b). Depending on the goal and scope of the study, the product system can be subdivided into a foreground system that relies on technology-specific primary data directly collected for the LCA study and into a background system, for which secondary datasets such as LCI databases are sufficient (Bjørn et al. 2018c).

The Life Cycle Impact Assessment (LCIA) translates the LCI into environmental impact scores. It aims at assessing the magnitude of each elementary flow to an impact on the environment. This is performed by means of impact category indicators for each impact category (Rosenbaum et al. 2018). LCA practitioners can draw on impact assessment methods that specify a number of category indicators with underlying specific characterization models and are available in common LCA software (Rosenbaum et al. 2018).

The final phase of the LCA is the Interpretation, which focuses on the identification of significant environmental issues in different stages of the life cycle of the product system. In order to assure the quality of the LCA, the Interpretation includes several checks regarding completeness, sensitivity and consistency (Hauschild et al. 2018).

5.2.2 Life Cycle Assessment of Electric Vehicles

When assessing electric vehicles, the variability of their environmental performance due to temporal, geographical and inter-individual aspects have to be considered. For a comprehensive overview of the Life Cycle Assessment methodology on electric mobility, please refer to (Cerdas et al. 2018). Figure 5.1 exemplarily shows the environmental impacts of using an electric vehicle as well as an internal combustion engine vehicle over its life cycle. Taking one impact category as an example, e.g. climate change, an electric vehicle has at the beginning of the use stage a higher environmental impact than a conventional vehicle. This is primarily due to the required raw materials for today's battery systems and the energy intensive battery production (Hawkins et al. 2013). The difference in environmental impact to the conventional vehicle at the end of the production stage (highlighted by ΔI) has to be compensated in the use stage. The point where this eventually happens is the environmental break-even point that is shown in the figure at x kilometers. However, the slope of the curves of the conventional and electric vehicles (α and β respectively) depends on the energy demand of vehicles and in the case of climate change as the chosen impact category on the carbon intensity of the energy used for fueling and charging the vehicles. The energy demand of vehicles regardless of their powertrain is composed of a share of energy demand for overcoming driving resistance forces (aerodynamic drag force, acceleration force, elevation force and rolling resistance), a share of energy

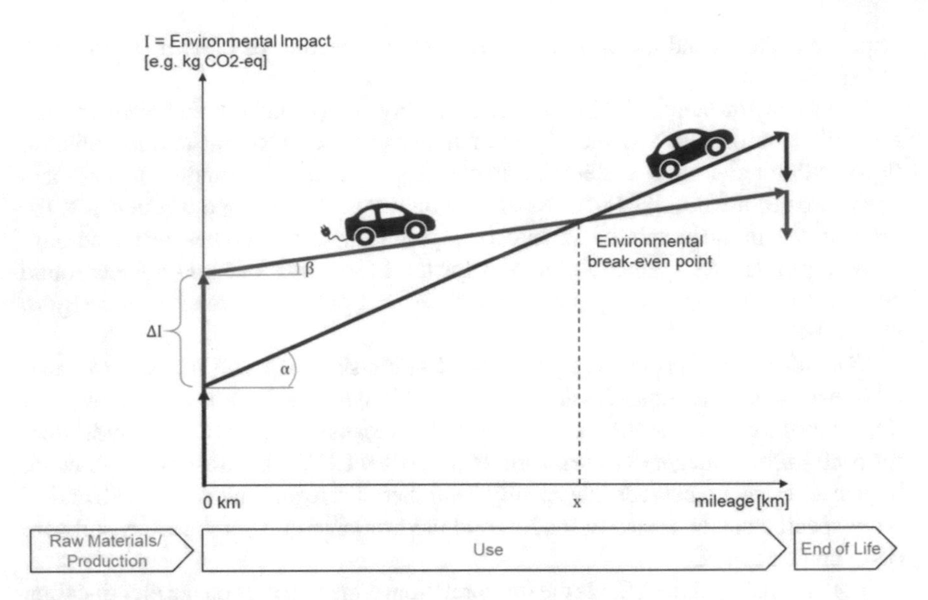

Fig. 5.1 Environmental impact of an electric vehicle over its life cycle in relation to a conventional vehicle

demand for powering auxiliaries (e.g. heating and cooling of passenger space), and losses (occurring due to inefficiencies in the energy conversion processes within the vehicle). After the end of the use stage, the vehicles undergo a series of end of life treatment processes that account for further emission or credits (e.g. through energy recovery) in the end-of-life stage.

The trade-offs between the different life cycle stages of electric and conventional vehicles make it apparent that electric vehicles are not per se better from an environmental perspective than conventional vehicles. In order to reach the environmental break-even point in the use stage, avoid problem shifting and provide an environmentally benign individual mobility with electric vehicles, it is necessary not only to engineer the vehicle but also to engineer the systems associated with the life cycle stages of electric vehicles.

5.2.3 Current Research Approaches and Related Work

Hawkins et al. (2012) and Nordelöf et al. (2014) provide comprehensive reviews on the Life Cycle Assessment of electric vehicles and conclude that LCA studies on electric and hybrid electric vehicles are often difficult to compare due to a number of factors such as selected impact categories, time scope, uncertainties and knowledge gaps in the supply chain. However, all LCA studies agree on the importance of the electricity mix during the use stage of electric vehicles.

In order to support the process of assessing the environmental impacts associated with electric vehicles and enhance the comparability of LCA studies, the European Commission supported the eLCAr project that adapted the guidelines of the ILCD Handbook, which provides general guidelines and a framework for LCA practitioners, to the specific case of electric vehicles (Del Duce et al. 2013). However, as electric mobility is still an emerging technology with a number of unresolved obstacles and technological issues (e.g. competing charging concepts in the use stage), most LCA studies focus on vehicle production (including specific components, e.g. the battery cell production) and treat the use stage on a more generic level. In this case, the use stage energy demand relies on static calculations based on standardized driving cycle tests. Yuan et al. (2017) point out that energy demand obtained from standardized driving cycles are far away from real-world drivetrain energy demand. They propose a physical-based statistical method for evaluating the energy demand of electric vehicles. Despite the high relevance of auxiliary energy demand in electric vehicles as shown in Chap. 2, Yuan et al. (2017) only focus on the energy demand related to overcoming the driving resistance forces in their method and neglect to address auxiliary energy demand in electric vehicles. Fiori et al. (2016) emphasize the need for a simple and accurate energy consumption model of electric vehicles in order to better predict remaining range, thus overcome range anxiety. Their proposed model computes drivetrain energy demand and energy demand for auxiliaries such as heating & cooling as well.

Egede et al. (2015) developed a framework to consider influencing factors in the LCA of electric vehicles. The presented framework combines internal factors (e.g. vehicle weight, production processes) with external factors in the use stage that cover aspects from the user (e.g. driving style and charging behavior), infrastructure (e.g. electricity mix and charging systems) and surrounding conditions (e.g. climate zone and topography). Li et al. (2016) reviewed possible factors influencing the energy demand of electric vehicles and conducted Design of Experiment studies to explore the statistical significance of the factors topography, infrastructure, traffic and climate as well as secondary effects by combination of these factors. They found that all of the factors have a significant impact on the energy demand of electric vehicles. de Cauwer et al. (2015) apply a multiple linear regression model to detect and quantify correlations between the kinematic parameters of the vehicle and its energy demand. Using the models, the vehicle's energy demand can be predicted on different levels of aggregation depending on the input parameters. Wang et al. (2017) analyzed the energy demand of electric vehicles in a field test. Main research focus was quantifying the dependency of energy demand on ambient temperature.

As the life cycle of an electric vehicle is inherently connected with the electricity generation and distribution system, the assessment of the use stage of electric vehicles needs to address the electricity generation adequately (Nordelöf et al. 2014). In this context, Ellingsen et al. (2016) discuss the life cycle emissions of different vehicle categories as a function of different electricity mixes in the use stage (electricity based on coal, natural gas and wind power in the use stage and additionally wind power in all life cycle stages).

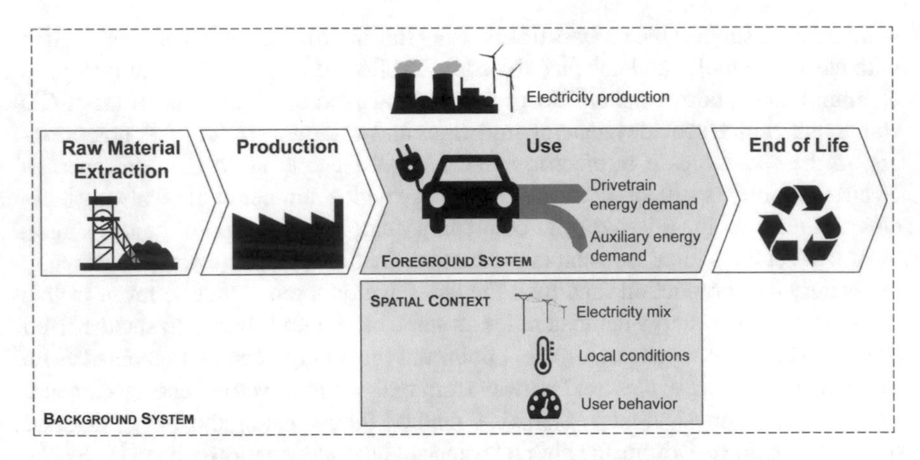

Fig. 5.2 System boundaries of the LCA study

5.3 Life Cycle Assessment of Electric Vehicles in Fleet Applications—A Case Study

The aim of this LCA study is to evaluate and illustrate the environmental impacts of two corporate electric vehicle fleets. The intended use of the results is to derive general recommendations targeting fleet owners, vehicle manufacturers and political decision makers. As the study is part of the project Fleets Go Green, other research modules of the project provided an input for the study and the main findings served as an input for other research modules.

The study is structured along two different scenarios. The first scenario describes the fleet operation at the local energy supplier, BS|Energy and covers 23 electric vehicles. The vehicle's use is explicitly limited for business purposes. The second scenario outlines a car-sharing scenario on the campus of the Technische Universität Braunschweig. The fleet is intended for university employees and consists of four electric vehicles. The main differences between the two scenarios are the fleet size and composition as well as the characteristics and resulting requirements on fleet usage.

Figure 5.2 illustrates the system boundaries and the respective foreground and background system of the LCA. The main focus of the LCA is on the modelling of the use stage with regard to a spatial context. It includes the observation of the drivetrain and auxiliary energy demand as well as the driving style. In particular, the interdependencies between real driving conditions in corporate fleets, local conditions and electricity mix are in the focus and represent the foreground system. The total energy demand per km and the breakdown of the total energy demand to component level are directly retrieved from the continuous measurement of each individual drive so that no other assumptions or literature values are considered during the modeling of use the stage.

Further life cycle stages such as the raw material extraction and acquisition, production, end-of-life processes and disposal of wastes constitute the background system. These life cycle stages are modeled according to the eLCAr Guidelines via generic vehicle models (Del Duce et al. 2013). The data used to set up the generic vehicle models is acquired from the bill of materials (BOM) of vehicle and subcomponent manufacturers. For confidentiality purposes, the data is aggregated to product group level, keeping all material and weight information constant. Recycling process models are generated based on bibliographical references and on industry data. The ecoinvent database (version 2.2) is applied to all life cycle stages in the background system to retrieve generic LCI data (Frischknecht et al. 2005). In addition, more than 300 data sets are modeled to constitute inventories representing for the special production processes in the automotive industry. These datasets are either built on basis of existing ecoinvent inventories (as a focused combination of them) or additional research has been performed in case of not existing datasets.

It is assumed that the vehicles are commercially used for a period of three years. Afterwards, the vehicles enter the private used car market. The scope of the LCA considers a distance of 150,000 km as the total lifetime of a vehicle. After this distance, the vehicles are assumed to reach their end of life and are treated in the required recycling processes.

The functional unit for both scenarios is defined as the transport of passengers and equipment for a distance of 150,000 km, which is the average driving distance of vehicles in corporate fleet operations (Hawkins et al. 2012). In order to allow an evaluation based on real world driving emissions, vehicles were equipped with metering devices. Therefore, additional information for driving conditions such as the number of passengers, weight of transported equipment and further specifications on the driving environment are not included in the functional unit.

The CML method 2001 is chosen for the Life Cycle Impact Assessment (Tukker et al. 2002). During impact assessment, five impact categories are evaluated. Besides climate change, acidification potential, eutrophication potential, photochemical ozone creation potential and ozone depletion potential are identified as relevant for the study. Table 5.1 lists the chosen impact categories, respective LCIA methods and units of measurements.

For the raw material extraction and production stage, all materials included in the vehicle are covered in the study. Due to the complexity of a vehicle and its subcomponents, the materials with a weight lower than 1 kg are mapped in LCA inventories with generic processes for each material category only. The generic datasets represent an average mixture of materials and processes. The allocation procedures and cut-offs considered in the ecoinvent inventories have been applied (e.g. average transport for raw materials, share of the required infrastructure). It is assumed that all components for the vehicle are fabricated within the same factory, even though this is not usually the case in reality. Hence, the transportation efforts caused between the production plants of the outsourced components, subparts and semi-finished materials to the final assembly plant as well as the delivery of the vehicles to the customer is not taken into account.

Table 5.1 Selected impact categories for the case study

Impact category	LCIA method	Unit
Acidification potential	CML 2001: acidification potential: average European	kg SO_2-Eq
Climate Change	CML 2001: climate change: lower limit of net GWP	kg CO_2-Eq
Eutrophication potential	CML 2001 w_o LT: eutrophication potential w_o LT: average European w_o LT	kg NO_x-Eq
Photochemical ozone creation potential	CML 2001: photochemical oxidation (summer smog): high NOx POCP	kg formed ozone
Ozone depletion potential	CML 2001: stratospheric ozone depletion: ODP 30a	g CFC-11-Eq

For the end of life stage, an input orientated allocation is chosen for recycling activities. This means that the benefit of using secondary materials are considered, but no further credit for the recycling is accounted for. During the production stage the cut-off materials are considered to be recycled.

5.4 Inventory Analysis of Electric Vehicles in Fleet Applications Within the Case Study

While the raw material extraction and acquisition, material production, and manufacturing as well as end of life stages are modeled via generic vehicle models, real world field test data is applied to the use stage calculations.

5.4.1 Generic Vehicle Models for Raw Material, Manufacturing and End of Life Stages

The approach for creating the generic vehicle models is to connect and to reuse existing product data from the bill of materials (BOM). In this context, vehicle's data (assemblies, components, materials, and substances) from the International Material Data System (IMDS) and from other internal data sources (e.g. related Product Data Management and Product Lifecycle Management systems) are integrated into a structured product model. In the following steps, the appropriate process data is anonymized, matched in accordance with the materials and the predefined material library and aggregated into larger assembly groups. The structure of the generic vehicle model is available including the assemblies shown in Fig. 5.3, which summarizes the assembly names and their share on the total vehicle weight for the battery electric vehicle (BEV) and thane internal combustion engine vehicle (ICEV).

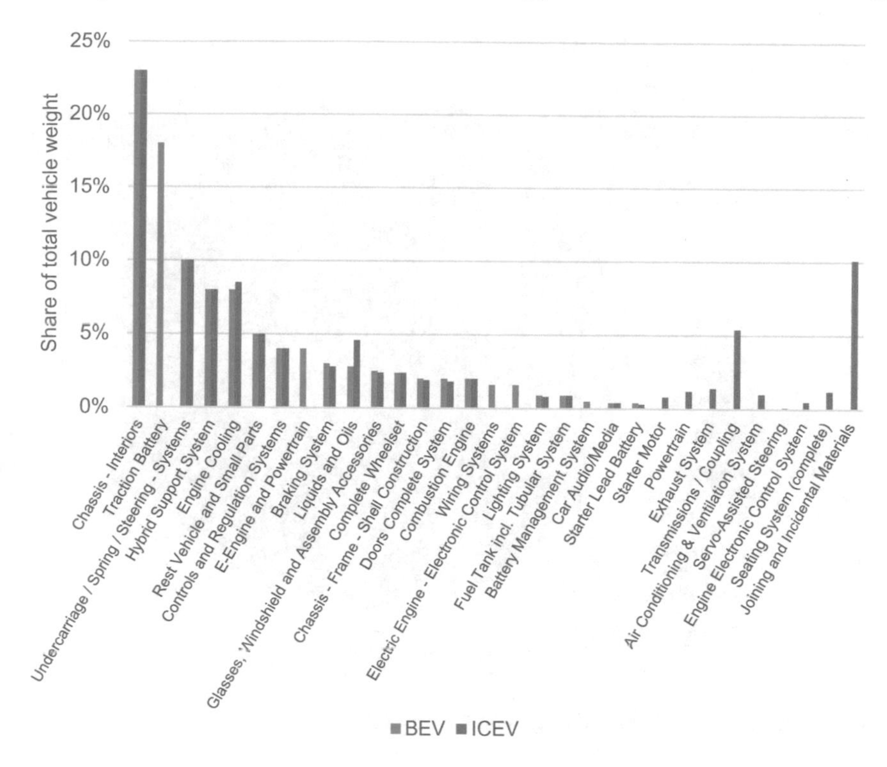

Fig. 5.3 Structure of the generic BEV and ICEV model

The material composition of the generic BEV is shown in Fig. 5.4. The graph defines material classes according to (VDA 231-106 1997). Steel and iron materials constitute the highest material fraction with a share of over 52% of the whole vehicle weight. The next largest fraction is polymers (by more than 18%). Lightweight metals (like aluminum and magnesium) represent another significant fraction with almost 12%. Electronics (6%) and fluids (3.5%) are relatively small material groups from the weight perspective.

For the inventory analysis of the raw material extraction and acquisition stage, the materials in each assembly of the BEV are matched to appropriate LCA datasets. The same material can be linked to different raw material production and manufacturing datasets for different assemblies or combination of materials, but within one assembly the datasets are the same for the same material. The assignment of LCA process inventories for the manufacturing of parts, components and assemblies happens similarly. For each assembly/material combination several LCA datasets can be assigned (e.g. sheet rolling, drilling, deep drawing and blanking), as they represent several steps within the production chain of BEV.

The LCIA of the generic BEV model builds the basis for the LCIA calculation of other vehicle types in the fleet. The weight of each assembly is extrapolated for each vehicle, according to the difference between the total weights. Based on these

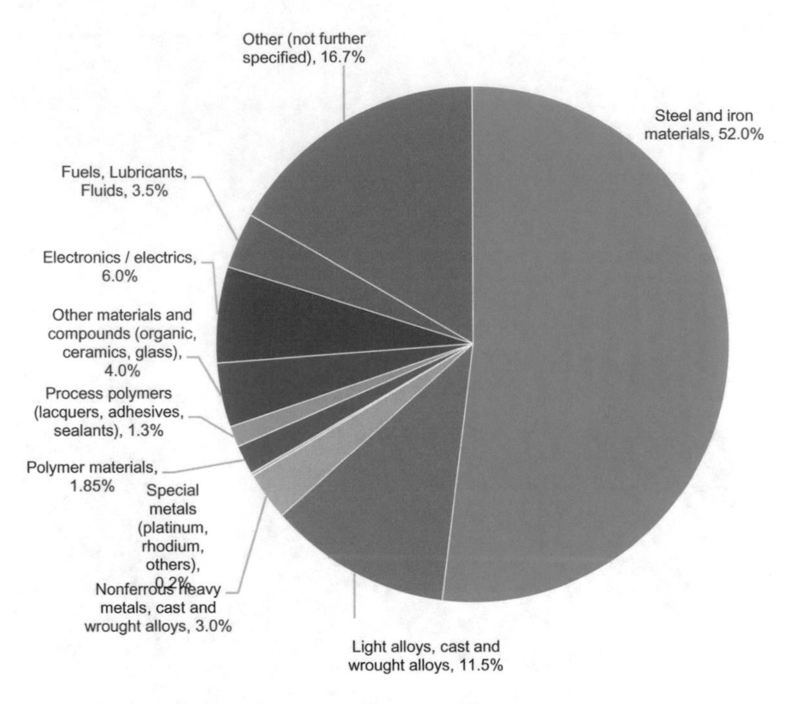

Fig. 5.4 Material composition of the generic BEV model

vehicle's specific weights per assembly, the LCIA is calculated by scaling the generic vehicle model's LCIA results per assembly. Therefore, the calculated LCIA for the generic BEV model is linearly scaled according to the weight distribution of each separate fleet vehicle. Summing up each vehicle's LCIA that belongs to the fleet gives the LCIA of the whole fleet.

5.4.2 *Use Stage*

The energy demand of the vehicles is directly acquired via a continuous measurement in a field test. Therefore, no other assumptions or literature values are considered for the calculation of the use stage environmental impacts. The energy demand is broken down to drivetrain energy demand and energy demand for auxiliaries such as air conditioning and heating.

The german electricity mix dataset from ecoinvent is used to model the impacts in the use stage. The lifetime of the traction battery is assumed to be 150,000 km (equal to the driving distance of the functional unit). Therefore, no additional traction battery is considered for the use stage of the vehicles. The lifetime of the tires is assumed to be 30,000 km. Thus, five sets of tires including the original ones are considered during the total lifetime of the vehicles. Other consumable materials required for service and maintenance (e.g. liquids and spare parts such as brake discs, wheels, and bumper) are not considered. For the generation and distribution of electricity, the allocation procedures and cut-offs considered in the ecoinvent inventories are applied (e.g. infrastructure, losses).

5.5 Impact Assessment of Electric Vehicles in Fleet Applications Within the Case Study

This section presents the results of the Life Cycle Impact Assessment for both scenarios. Sensitivity analysis are conducted in order to test the reliability of the models and the influence of varying factors on the impact assessment results. The results of the sensitivity analysis show the influence of the electricity mix, ambient temperature and driving patterns on the environmental impacts of BEV in fleet operations.

5.5.1 Impact Assessment of the Corporate Fleet Scenario

Figure 5.5 demonstrates the environmental impact assessment results of the corporate fleet scenario at BS|Energy for selected impact assessment categories. EU-25 "normalization" factors in line with CML 2001–11/20106 are adopted in order to normalize the results. As depicted in the figure, the use stage is the most dominant contributing factor across the selected impact assessment categories. Regarding climate change, the biggest contribution arises by the energy demand of electric vehicles in the use stage, and consequently by the generation and distribution of electricity. The second and third dominant contributors are raw material extraction and the production stage.

The car body and lithium-ion battery have the highest impacts for both raw material and production stages (car body 16% in raw material extraction and 32% in the production stage, battery 14% in raw material extraction and 11% in production stage). Eutrophication potential is dominated by raw material production. The most influencing contribution after use stage for eutrophication potential is the production of chassis-suspension-steering systems with 24%. The car body and battery production have also important contributions with 12 and 14%. Car body and interior body production are dominant for acidification potential and photochemical ozone creation potential.

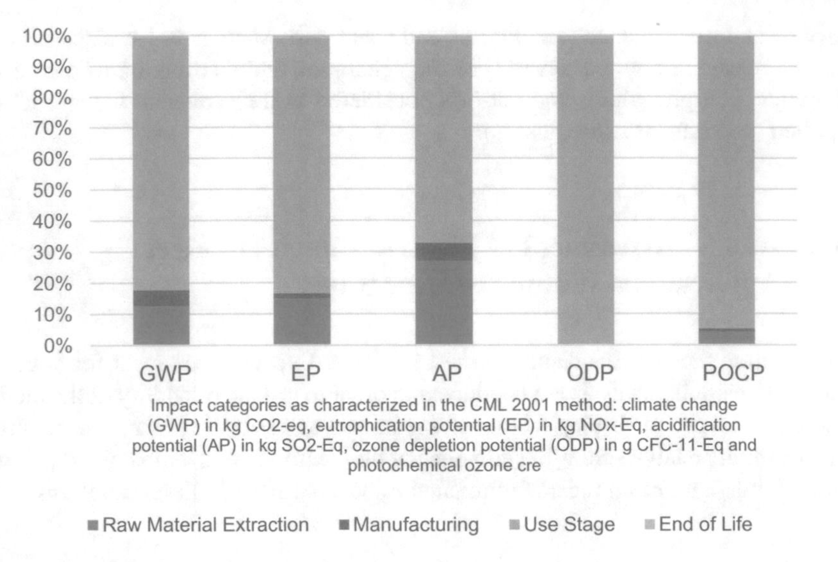

Fig. 5.5 Contribution analysis of the environmental impacts of the corporate fleet scenario

5.5.2 Impact Assessment of the Car Sharing Scenario

The environmental impact assessment results of the car sharing scenario at Technische Universität Braunschweig is displayed in Fig. 5.6. The biggest contribution to climate change results from the use stage, where all the energy generation related emissions are accounted for. Here it should be noted that maintenance is not considered in the use stage evaluations. The use stage has the highest contribution to eutrophication potential. Acidification potential is mainly influenced by raw material extraction and the use stage. Raw material and production stages are relevant due to the car body and interior body productions for acidification and photochemical ozone creation potential. The highest impact on climate change belongs to the electricity generation process. Acidification potential is dominated by the raw material and production stages of interior body and car body processes.

5.5.3 Influence of Different Electricity Mixes

The environmental impact assessment of both scenarios shows that the use stage has the highest contribution on climate change. In order to explore the influence of the electricity mix on the environmental impact in the use stage, a sensitivity analysis is conducted on the corporate fleet scenario at BS|Energy.

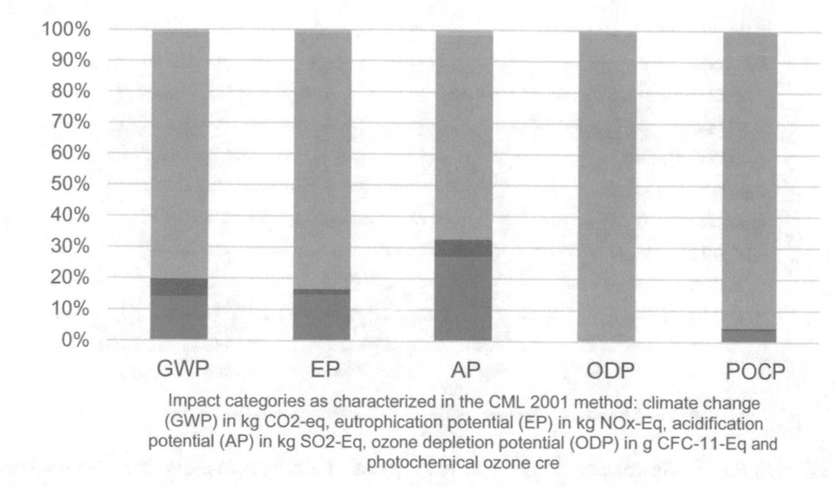

Fig. 5.6 Contribution analysis of the environmental impacts of the car sharing scenario

Table 5.2 Overview of the different electricity mix scenarios for the sensitivity analysis

Fleet scenarios	Abbreviation	Energy source	Dataset
Reference fleet of BS|Energy	BS_Ref	Diesel/petrol	Operation, passenger car, diesel, EURO5 [RER] Transport, passenger car, petrol, EURO5 [RER]
Electric vehicle fleet of BS|Energy	BS_EV	German electricity mix	Electricity, production mix [DE]
Electric vehicle fleet solar power	BS_Solar	Solar energy	Electricity, production mix photovoltaic, at plant [DE]
Electric vehicle fleet wind power	BS_Wind	Wind energy	Electricity, at wind power plant [RER]
Electric vehicle fleet coal power	BS_Coal	Coal power	Electricity, hard coal, at power plant [DE]

Three additional energy sources served as the basis for the sensitivity analysis: a solar energy fleet (BS_Solar), a wind energy fleet (BS_Wind) and a coal power fleet (BS_Coal). The baseline electric vehicle fleet (BS_EV) uses the german electricity mix. Additionally, the environmental impacts of the reference fleet with solely conventional vehicles are calculated to illustrate the impact of the energy source in relation to the electric vehicle fleets. The reference fleet consists of 19 vehicles; six of them are diesel and 13 petrol vehicles. Table 5.2 compiles the different electricity mix scenarios and the respective datasets from ecoinvent.

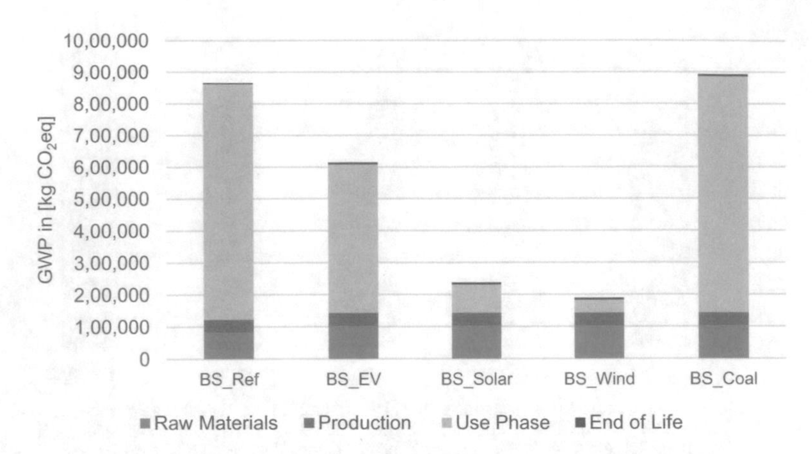

Fig. 5.7 Impact of different electricity mixes to the global warming potential in the corporate fleet scenario

Figure 5.7 displays the global warming potential of the reference fleet and the electrified fleet with different electricity mix. The results reflect that each of the chosen electricity sources has a different carbon intensity. The graph clearly underlines how the environmental impact of the fleets is dependent on the electricity mix. Even though the conventional fleet has a lower impact in the production stage, it has in total a higher global warming potential than the electric vehicle fleet. This difference gets even higher when electricity from renewable sources is used to power the electric fleet. On the contrary, when the electric vehicle fleet is charged with electricity from carbon-intensive fossil sources, coal in this case, it has a higher GWP than the reference fleet with ICEV's. These results highlight that the environmental performance of electric vehicle fleets is strongly dependent on the use stage electricity mix for charging.

However, vehicle size and number of vehicles can be discussed when comparing the conventional fleet to the electrified. The conventional reference fleet of BS|Energy consists of 19 vehicles, whereas the electric vehicle consists of 23. On the other side, vehicles in the electric vehicle fleet are small size vehicles, whereas the conventional fleet has 13 small size vehicles and six panel vans, which are used for transportation purposes. The impact of vehicle sizes to the overall energy demand of electric/conventional fleet can be a focus in a future research project.

5.5.4 Influence of Ambient Temperature

Capturing real world field test data enables the inquiry of drivetrain and auxiliary energy demand for each vehicle. Following section explores the influence of the intensity of heating and cooling of the passenger space to the environmental impact.

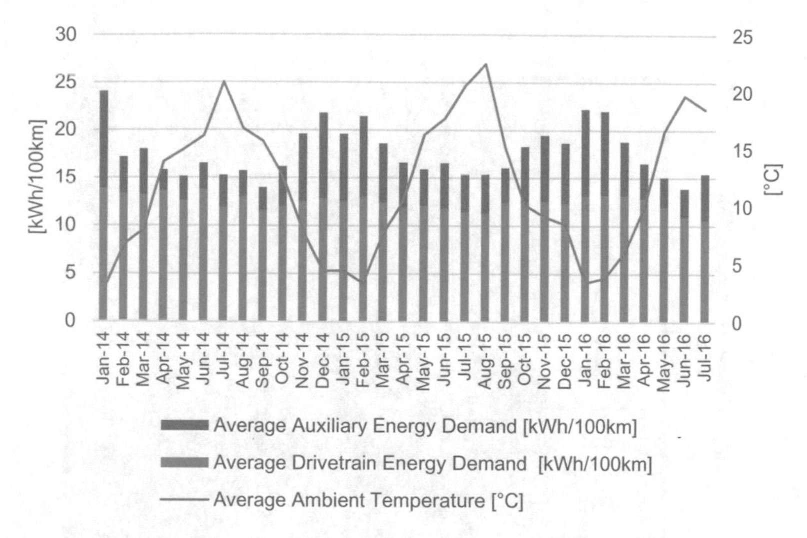

Fig. 5.8 Energy demand of electric vehicles in relation to ambient temperature

Figure 5.8 illustrates the drivetrain and auxiliary energy demand of one representative small size fleet vehicle over the duration of the field test that lasted for 2.5 years. The graph additionally displays the average ambient temperature and correlates therefore energy demand of electric vehicles with ambient temperatures.

The records point out a trend towards lower total energy demand during warmer summer months and higher total energy demand in colder winter months. According to the breakdown of auxiliary and drivetrain energy demand, the share of auxiliary energy demand mainly creates this fluctuation in the graph. The auxiliary energy demand can be linked with the intensity of air conditioning and heating of the passenger space (required to maintain the preferred indoor temperature). While drivetrain energy demand remains on an approximately same level over the duration of the field test, heating and cooling accounts for 24 and 38% of total energy demand in winter and summer months respectively. Air conditioning in summer months proved to be less energy intensive in the field test that took place in Braunschweig, northern Germany. Depending on the local climate, the absolute amount of the energy demand of heating and cooling and its share to total energy demand could differ from the presented result in Fig. 5.8.

Figure 5.8 displays that different ambient temperatures, thus local conditions have a great influence on the total energy demand of electric vehicles. Hence, ambient temperature is an important aspect not only when calculating the range of a BEV but also when assessing the environmental impacts of electric vehicle operations.

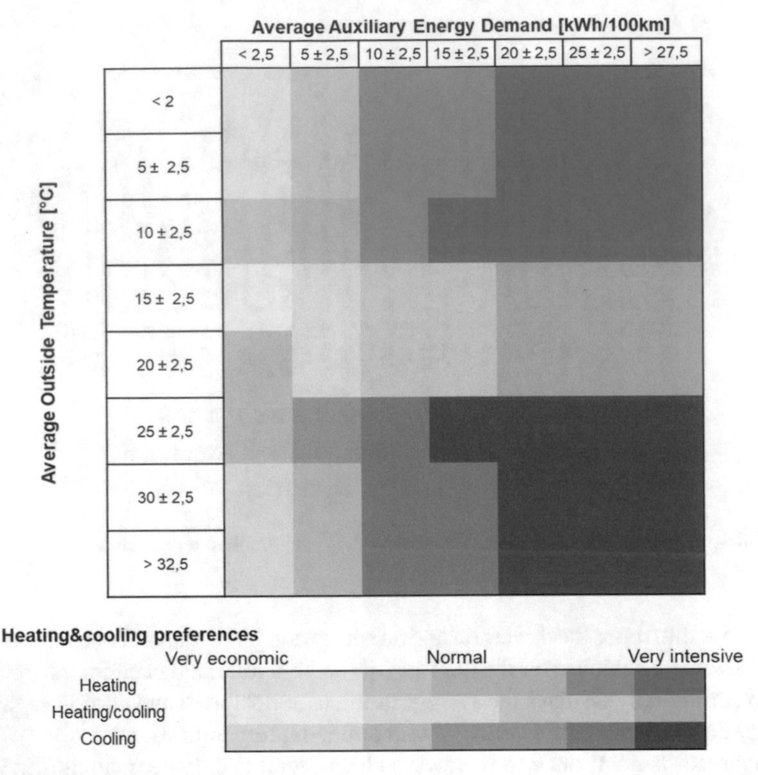

Fig. 5.9 Linking heating cooling preferences with ambient temperature and auxiliary energy demand

5.5.5 *Influence of Driving Parameters*

The data acquisition during the operation of the vehicles also allow for assessing the driving environment, driving style, ambient temperature and personal heating and cooling preferences. The aim of this section is to explore the influence of driving parameters on the overall environmental impact of fleets. First, the average drivetrain energy demand is derived for different driving environments (autobahn, rural, city or mix) and driving styles (economic and sportive). Secondly, the energy demand for auxiliaries is mapped as a function of the driver's heating and cooling preferences and ambient temperature. The clusters displaying average ambient temperatures, describing the use intensity of auxiliaries and linking them to auxiliary energy demand are shown in Fig. 5.9. The data in the graph is based on a small size fleet vehicle. The heat map shows the auxiliary energy demand for a given ambient temperature and heating and cooling preference of the driver.

Later on, different user profiles are created by coupling the driving environment/style parameters with the use intensity of auxiliaries. Figure 5.10 displays a selection of different user profiles in this context. The first criterion makes a spa-

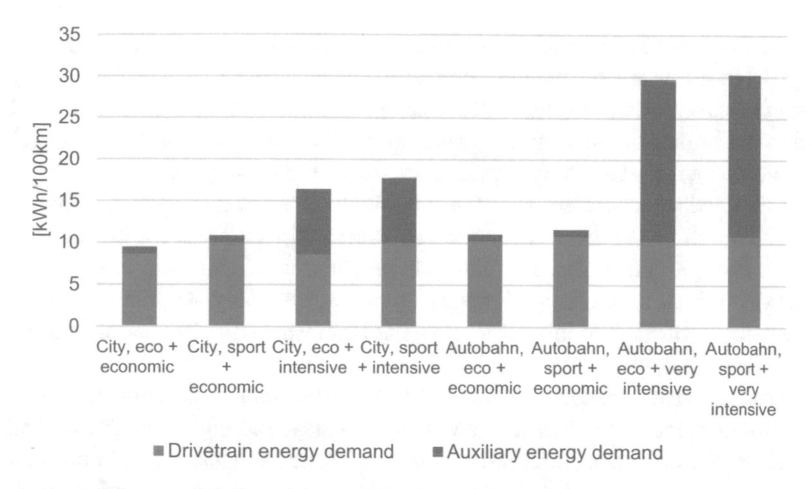

Fig. 5.10 Impacts of driving parameters on the energy demand

tial differentiation of usage scenarios and describes the driving environment (city or autobahn). The second criterion focuses on the individual the driving style (eco or sportive) of the driver. A sportive driving style is characterized by higher acceleration and deceleration rates that were acquired from the field measurement. The third criterion describes the use intensity of auxiliaries from Fig. 5.9. For example, a city driver with economic driving style, who has economic heating and cooling preferences, has 12.72 kWh energy demand over 100 km, whereas an autobahn driver with a sportive driving style and very intensive auxiliary use intensity consumes 40.62 kWh per 100 km. The energy demand for auxiliaries exceeds in this case the drivetrain energy demand by the factor two. This implies a broad range of results that is often simplified to one mean value in LCA calculations. Moreover, Figs. 5.9 and 5.10 highlight that the individual driving style and heating and cooling preferences have a significant impact on the energy demand, thus environmental impact of electric vehicles. The presented data is valid for a small size vehicle in the fleet, but the methodology is applicable to other vehicle types as well.

These comparisons underline the dynamic nature of the energy demand of electric vehicles and respective environmental impacts. While the energy demand of electric vehicles is highly dependent on the use intensity of auxiliaries and driving patterns, it is in most LCA studies underestimated.

5.6 Summary and Outlook

Increasing the share of electric vehicles in road transport, especially in corporate vehicle fleets, is a promising way to decrease transport-related emissions. However, the environmental performance of electric vehicle fleets is dependent on background systems. Factors like the electricity mix, local climate and user behavior have an indisputable influence on the environmental impacts. In order to achieve environmental conscious electric vehicle fleet management, it is of vital importance to keep in mind the bigger picture and address improvement measures not only on vehicle level but also with regard to energy supply and charging infrastructure.

A Life Cycle Assessment study of electric vehicles in fleet applications is conducted. Two different fleet scenarios are evaluated, representing different usage patterns. First, a corporate fleet management scenario at the local energy supplier BS|Energy, which consists of 23 electric vehicles. Secondly, a car sharing scenario on the campus of the Technische Universität Braunschweig, which consists of four electric vehicles.

In both scenarios, generic vehicle models are developed and applied to evaluate the environmental impacts of the raw material extraction and acquisition, material production, manufacturing and end of life stages. The energy demand of the vehicles in the use stage is directly acquired via a continuous measurement in a field test. The acquired data differentiates between drivetrain energy demand and energy demand for auxiliaries such as air conditioning and heating.

The energy demand of the vehicles accounts for the highest global warming potential. As shown in the sensitivity analysis, the electricity mix has a great influence on the overall results. Hence, the generation of electricity plays an important role. In order to ensure an environmentally benign fleet operation, the source of electricity for charging cannot be underestimated. Efforts to increase the share of renewable energies in the electricity mix would support the emission reduction targets via electric mobility.

However, the total energy demand of an electric vehicle not only consists of the drivetrain energy demand but also of a significant share for auxiliaries such as heating and cooling. The sensitivity analysis revealed the influence of ambient temperatures and respectively of the local climate on the auxiliary energy demand of electric vehicles. The obtained results show that the electric vehicle fleet has a higher energy demand in winter moths than in summer months. It should be noted that the vehicles have been used in a moderate climate in Braunschweig, Germany that has no extreme cold or hot weather periods. Different climate zones may have a varying influence on the share of heating and cooling of the passenger space of electric vehicles. Based on the driver's indoor temperature preference, the auxiliary energy demand of vehicles varies dramatically. Furthermore, the influence of the driving style (economic or sportive) and driving environment (autobahn, rural, city or mix) cannot be neglected either. The sensitivity analysis underlines the importance of driving patterns on the results of environmental assessment.

In order to address the high importance of auxiliary energy demand from a technical perspective, insulation improvements in electric vehicles or improved design aspects for a better internal thermo-management can be further be investigated by vehicle manufacturers. Fleet owners may concentrate on sensitizing employees to the economic driving options and consider the impacts of different driving routines in daily business.

The overall results show that the environmental performance of electric vehicle fleets underlies numerous influencing factors in the use stage. Not only the electricity mix but also driver's related aspects such as indoor temperature preference, driving environment and driving style have a dramatic influence on the environmental impacts of using electric vehicles in fleet applications. The results indicate that in order to have a robust environmental performance evaluation, the intensity of auxiliary usage and personal driving preferences, should be an integral part of LCA studies on electric vehicles.

Acknowledgements The authors express their gratitude to the German Federal Ministry for the Environment, Nature Conservation, Building and Nuclear Safety for supporting the project "Fleets Go Green—Integrated analysis and evaluation of the environmental performance of electric and plugin-hybrid vehicles in everyday usage on the example of fleet operations" under the reference 16EM1041.

References

Del Duce A, Egede P, Öhlschläger G, Dettmer T, Althaus H-J, Bütler T, Szczechowicz E (2013) eLCAr: guidelines for the LCA of electric vehicles

Bjørn A, Laurent A, Owsianiak M, Olsen SI (2018a) Goal Definition. In: Hauschild MZ, Rosenbaum RK, Olsen SI (eds) Life cycle assessment. Springer International Publishing, Cham, pp 67–74

Bjørn A, Moltesen A, Laurent A, Owsianiak M, Corona A, Birkved M, Hauschild MZ (2018b) Life cycle inventory analysis. In: Hauschild MZ, Rosenbaum RK, Olsen SI (eds) Life cycle assessment. Springer International Publishing, Cham, pp 117–165

Bjørn A, Owsianiak M, Laurent A, Olsen SI, Corona A, Hauschild MZ (2018c) Scope Definition. In: Hauschild MZ, Rosenbaum RK, Olsen SI (eds) Life cycle assessment. Springer International Publishing, Cham, pp 75–116

Cerdas F, Egede P, Herrmann C (2018) LCA of Electromobility. In: Hauschild MZ, Rosenbaum RK, Olsen SI (eds) Life cycle assessment: theory and practice. Springer International Publishing, Cham, pp 669–693

de Cauwer C, van Mierlo J, Coosemans T (2015) Energy consumption prediction for electric vehicles based on real-world data. Energies 8:8573–8593. https://doi.org/10.3390/en8088573

Egede P, Dettmer T, Herrmann C, Kara S (2015) Life cycle assessment of electric vehicles—a framework to consider influencing factors. Procedia CIRP 29:233–238. https://doi.org/10.1016/j.procir.2015.02.185

Ellingsen LA-W, Singh B, Strømman AH (2016) The size and range effect: lifecycle greenhouse gas emissions of electric vehicles. Environ Res Lett 11:54010

Fiori C, Ahn K, Rakha H (2016) Power-based electric vehicle energy consumption model: model development and validation. Appl Energy 168:257–268

Frischknecht R, Jungbluth N, Althaus H-J, Doka G, Dones R, Heck T, Hellweg S, Hischier R, Nemecek T, Rebitzer G, Spielmann M (2005) The ecoinvent database: overview and methodological framework (7 pp). Int J Life Cycle Assess 10:3–9. https://doi.org/10.1065/lca2004.10.181.1

Hauschild MZ, Bonou A, Olsen SI (2018) Life Cycle Interpretation. In: Hauschild MZ, Rosenbaum RK, Olsen SI (eds) Life cycle assessment. Springer International Publishing, Cham, pp 323–334

Hawkins TR, Gausen OM, Strømman AH (2012) Environmental impacts of hybrid and electric vehicles—a review. Int J Life Cycle Assess 17:997–1014. https://doi.org/10.1007/s11367-012-0440-9

Hawkins TR, Singh B, Majeau-Bettez G, Strømman AH (2013) Comparative environmental life cycle assessment of conventional and electric vehicles. J Ind Ecol 17:53–64. https://doi.org/10.1111/j.1530-9290.2012.00532.x
IEA (2017) CO2 emissions from fuel combustion: overview. https://www.iea.org/publications/freepublications/publication/CO2EmissionsFromFuelCombustion2017Overview.pdf
ISO 14040:2006 (2006) Environmental management—life cycle assessment—principles and framework
Kasten P, Zimmer W, Leppler S (2011) CO_2-Minderungspotenziale durch den Einsatz von elektrischen Fahrzeugen in Dienstwagenflotten: Ergebnisbericht im Rahmen des Projektes „Future Fleet" AP 2.7. http://www.oeko.de/oekodoc/1343/2011-027-de.pdf
KBA (2015) Neuzulassungen von Pkw im Jahr 2015 nach privaten und gewerblichen Haltern. https://www.kba.de/DE/Statistik/Fahrzeuge/Neuzulassungen/Halter/2015/2015_n_halter_dusl.html
Li W, Stanula P, Egede P, Kara S, Herrmann C (2016) Determining the main factors influencing the energy consumption of electric vehicles in the usage phase. Procedia CIRP 48:352–357. https://doi.org/10.1016/j.procir.2016.03.014
Nordelöf A, Messagie M, Tillman A-M, Ljunggren Söderman M, van Mierlo J (2014) Environmental impacts of hybrid, plug-in hybrid, and battery electric vehicles–what can we learn from life cycle assessment? Int J Life Cycle Assess 19:1866–1890. https://doi.org/10.1007/s11367-014-0788-0
Rosenbaum RK, Hauschild MZ, Boulay A-M, Fantke P, Laurent A, Núñez M, Vieira M (2018) Life cycle impact assessment. In: Hauschild MZ, Rosenbaum RK, Olsen SI (eds) Life cycle assessment: theory and practice. Springer International Publishing, Cham, pp 167–270
Tukker A, de Bruijn H, van Duin R, Huijbregts MAJ, Guinee JB, Gorree M, Heijungs R, Huppes G, Kleijn R, de Koning A, van Oers L, Wegener Sleeswijk A, Suh S, Udo de Haes HA (2002) Handbook on life cycle assessment, vol 7. Springer, Netherlands, Dordrecht
VDA 231-106 (1997) Werkstoff-Klassifizierung im Kraftfahrzeugbau – Aufbau und Nomenklatur. VDA, Berlin
Wang J-b, Liu K, Yamamoto T, Morikawa T (2017) Improving estimation accuracy for electric vehicle energy consumption considering the effects of ambient temperature. Energy Procedia 105:2904–2909. https://doi.org/10.1016/j.egypro.2017.03.655
Yuan X, Zhang C, Hong G, Huang X, Li L (2017) Method for evaluating the real-world driving energy consumptions of electric vehicles. Energy 141:1955–1968. https://doi.org/10.1016/j.energy.2017.11.134

Chapter 6
Workshop Based Decision Support Methodology for Integrating Electric Vehicles into Corporate Fleets

Mark Stephan Mennenga, Antal Dér and Christoph Herrmann

6.1 Introduction and Motivation

Triggered by the discussion on climate change, the scarcity of fossil resources and the resulting intensified legislation and consumer awareness for sustainability, the automotive sector is changing (Kreyenberg 2016). The demand for a more sustainable and energy efficient mobility forces car manufacturers to incorporate alternatively powered vehicles into their product portfolio. In consequence, they have introduced a number of battery electric vehicles (BEVs) and plug-in hybrid electric vehicles (PHEVs) to the automotive market in recent years. For example, the global electric vehicle stock has already reached the threshold of 1 million EVs due to the definition of ambitious policies. In countries like France, Germany, Korea, Sweden, UK and India electric car sales grew over 75% between 2014 and 2015 (IEA 2016).

Corporate fleet owners are among the early adopters using the electrification trend to contribute to internal environmental targets, improve image and elaborate new business models. Due to the overall rising costs of fossil fuels and an increasing demand for and subsidisation of sustainable and energy efficient mobility, alternatively powered vehicles are more and more of concern. In fact, fleet managers see high potentials in a structural change in their fleets with regard to alternatively powered vehicles and assume their fleets to change to mixed fleets in the following years, consisting of conventional, electric and hybrid electric vehicles (LeaseTrend AG 2012; Vogel 2016).

Not least, the high economic and environmental relevance give corporate vehicle fleets a special role in the transfer of the automotive sector to more electrified solutions. Fleets may account for the second largest cost factor besides personnel

M. S. Mennenga (✉) · A. Dér · C. Herrmann
Institute of Machine Tools and Production Technology (IWF), Technische Universität Braunschweig, Braunschweig, Germany
e-mail: m.mennenga@tu-braunschweig.de

C. Herrmann et al. (eds.), *Fleets Go Green*, Sustainable Production, Life Cycle Engineering and Management,
https://doi.org/10.1007/978-3-319-72724-0_6

costs. They also account for a major share of the corporate environmental impact which results from e.g. the high kilometres travelled and high shares in annual new car registrations, e.g. annual new car registrations in corporate applications rose in Germany from 60% in 2008 to 66% in 2015 (Kasten et al. 2011; Statista 2016b; Kraftfahrt-Bundesamt 2015). Furthermore, they gain an important role transferring alternatively powered vehicles and new technologies to the used car market. Thus, an electrification of the fleet sector has not only a high impact but also reaches broad sections of the society (Mennenga 2014; Statista 2016b).

The operation of a vehicle fleet with one or more battery electric vehicles and plug-in hybrid electric vehicles lead to a new increased complexity in the planning phase of corporate fleets. Electric vehicles cannot readily substitute conventional vehicles with an internal combustion engine. Factors like the limited range, long charging times, different cost structures and indirect emissions through the electricity generation need to be addressed adequately. Here, it is essential to take a life cycle perspective, as only then the economic and environmental advantages and disadvantages of electric vehicles might be well understood (Mennenga 2014). For example, BEVs are connected with high purchase costs and emit no direct Greenhouse Gas (GHG) emissions during their use phase (other than the particles emitted due to break and tire wear etc.). However, due to the indirect emissions resulting from electricity production, the technology is still far from a "zero emissions" technology (Cerdas et al. 2016; Ellingsen et al. 2016). A key component of a life cycle oriented strategic fleet planning is thus the determination of an optimal mix of electric vehicles, plug-in hybrids and conventional vehicles, taking into account functional, economic and environmental aspects (Schrempf 2015). Ideally, the result of the fleet planning is a vehicle fleet that effectively performs its tasks at minimal costs, as well as risks, and environmental impact.

Fleet planning and management is often seen as an overhead task, as it is not the core business of most companies that operate fleets. Therefore, in order to cut costs, companies tend to outsource fleet planning activities or buy fleet planning competences (Schrempf 2015). As a result, the need for a decision support methodology is justified that supports fleet owners as well as fleet management companies in planning fleets that fulfil functional requirements under consideration of minimal costs, and environmental impacts. In the following, we present general implications of decision making in corporate fleet planning and derive requirements for a decision support methodology. This follows the evaluation of the state of research. On this basis, we present a workshop-based concept for the decision support of life cycle oriented vehicle fleet planning and apply it within a case study.

6.2 Decision Making in Corporate Fleet Planning

Decision making in corporate fleet planning aims at the identification, analysis and evaluation of alternatives for the composition of the fleet vehicle stock and the organization of the fleet vehicle deployment (Mennenga 2014). The fleet planning process

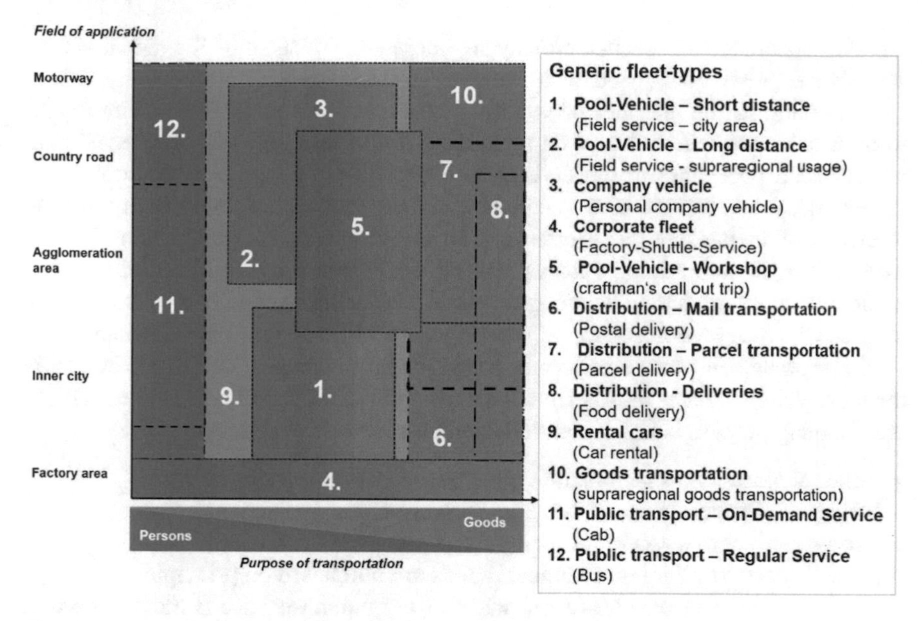

Fig. 6.1 Generic fleet types (Mennenga 2014)

is part of the fleet management that covers strategic, tactic and operational activities. To this end, there is a need for a deep understanding of the usage scenarios of targeted fleet operations. In general, corporate fleets are very heterogeneous and vary by different operational characteristics of the vehicles, owners, fleet sizes and their role or position within the company (Mennenga 2014; Gnann et al. 2012; Kerler 2003; Nesbitt and Sperling 2001). In order to enable a targeted fleet planning, a detailed classification of fleet types enables the specification of the requirements of potential fleets, such as required range or personnel and material vehicle capacity. Figure 6.1 shows twelve generic fleet types by means of the criteria area of use and purpose of transport. As elaborated by Mennenga (2014) the generic fleet types can be the basis to describe fleet requirements with five characteristic criteria groups: order volume, operational purpose, area of use, characteristic distances and possible stopovers.

In order to analyse decision processes in fleet planning, qualitative expert interviews among fleet planners in different industrial sectors with small fleets of less than 20 vehicles have been conducted and the results were evaluated according to the procedure of Kaiser (2014). The target group of the expert interviews resulted from the assumption that especially small and medium sized enterprises (SMEs) have a need for support with regard to the integration of alternatively powered vehicles into their corporate fleets. Typically, supporting organisations, such as the association of brand independent fleet management corporations in Germany (VMF) have a special focus on larger fleets with more than 20 vehicles (VMF 2017). However, the analysis of the current fleet market shows that up to more than three-fourth of the 4.5 million vehicles in corporate fleets in Germany are operated in fleets of SMEs with less than 20 vehicles (VMF 2017; Statista 2016a).

According to the interviewed experts, decision processes in SMEs with regard to their fleets are seldom based on systematic processes. Typically, this leads to an unstructured procedure in the division and execution of fleet management tasks. In general, the relevance of a systematic vehicle fleet management rises with the relevance of the fleet within the company, i.e. due to the relative capital commitment. The tasks split into strategic fleet planning, and more operational activities such as fleet support service or fleet organization. The central scope of duties typically includes the everyday fleet support service such as fleet service or maintenance.

Strategic decisions are a major challenge for fleet managers within SMEs as it is the task, which deviates mostly from the daily and more operational business. Often, the planning process is less structured and characterized as follows:

- minor preparation of the planning process
- little pre-information about the supply situation and maintenance costs
- strongly subjective opinion formation and influence on the decision
- partially certain vehicles or vehicle brands are purchased on principle
- only a few criteria are considered whereof the purchase price is most important and the total cost of ownership (TCO) are seldom determined; often, the term TCO is unknown
- the primary criterion for the decision is the functionality of the vehicles, followed by the acquisition costs, environmental criteria play a minor part
- maintenance costs are classified as less relevant.

In contrast, the planning process of a more systematic fleet management has the following characteristics:

- defined planning approach
- regular information search about the market situation and maintenance costs, irrespective of imminent new acquisition
- decision-making is based on hard facts, which are considered and compared within each new purchase
- key figures are analysed at regular intervals during everyday operation
- TCO or similar cost models are considered
- the primary criterion for the decision is the functionality of the vehicles, followed by the acquisition costs, environmental criteria are not in the focus
- awareness of the influence of maintenance costs.

A challenge of a systematic fleet management is the data acquisition of the vehicle fleet operation and its evaluation. In daily practice, often costs and daily or yearly mileage are recorded. Due to difficulties regarding privacy protection GPS monitoring is challenging. Normally, there is no evaluation of movement profiles or the driven distances of single fleet tasks. With respect to electric vehicles this information makes it difficult to understand if electric vehicles are suited for a specific fleet operation (compared to conventional cars).

The requirements for a decision support for the integration of alternatively powered vehicles in corporate fleet planning are elaborated by Mennenga (2014):

- **Consideration of conventionally <u>and</u> alternatively powered vehicles**: With respect to the composition of the fleet vehicle stock and the organization of the fleet vehicle deployment, conventionally and alternatively powered vehicle concepts need to be considered for the analysis and evaluation: This makes the configuration of mixed fleets possible.
- **Integrated consideration of the strategic <u>and</u> operational planning**: An integrated consideration of the fleet stock and fleet deployment planning is necessary because fleet vehicles are purchased for a few years and deployed in daily business. Thus, decisions in the strategic fleet planning are effective in the short-, mid- and long-term.
- **Process support**: Guidelines that correlate with the steps of the general planning process should support a fleet planner.
- **Reference to the decision situation of a fleet planner**: The decision support must be so generic that the integration of different fleet scenarios is possible. However, the decision support system must have a reference to the specifics of a certain fleet application.
- **Life cycle oriented evaluation of functional, economic <u>and</u> environmental targets**: A fleet planner must evaluate fleet alternatives against functional, economic and ecological criteria. In this way, conflicts of goals can arise. However, it is essential to take a life cycle perspective so that economic and environmental advantages and disadvantages can be well understood. From an economic point of view, the consideration of the TCO is crucial. From an environmental point of view, the evaluation must include life cycle emissions in different impact categories (such as climate change).
- **Capture dynamic effects**: The economic and environmental impact of electric fleet vehicles is influenced by many diverse factors like the weather, driving style, vehicle weight due to loading, the number of persons as well as the source of the energy used to power the vehicle, as reviewed by Egede et al. (2015). These factors affect the range of the fleet vehicles in daily business and thus their availability. Consequently, it is necessary to capture dynamic effects.
- **Easy handling and practicality**: Finally, a decision support must be easy to handle and to integrate into existing corporate processes. Especially for SMEs auxiliary material and visual aids are helpful to overcome the inherent complexity of fleet planning with alternatively powered vehicles.

Based on these requirements the state of research can be analysed.

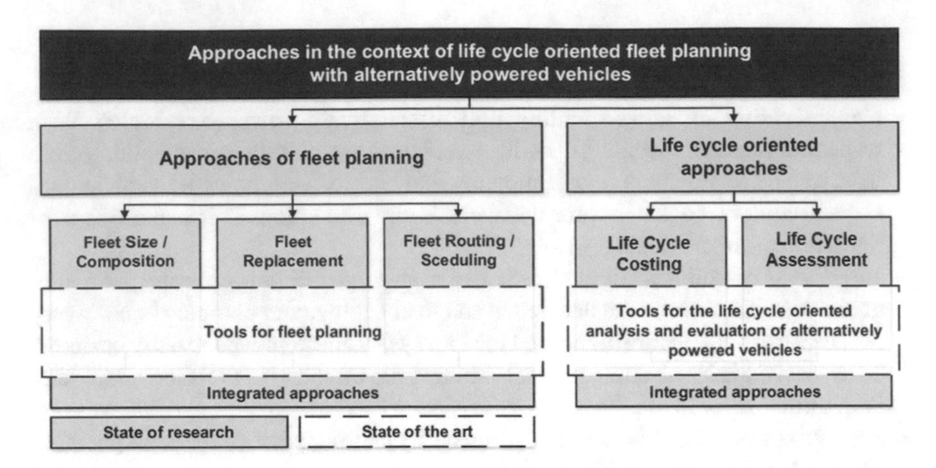

Fig. 6.2 Categorization of relevant research approaches (Mennenga 2014)

6.3 Related Works and Current Research Approaches

Approaches for the planning of fleets with conventional vehicles are numerous in literature. These approaches mainly originate from the discipline of the Operations Research (Mennenga 2014). The primary focus is on the determination of the:

– optimal fleet composition and size (Fleet Composition/Size problems): what is the optimal size of the fleet and what kind of vehicles are required to perform a given task,
– optimal replacement schedule (Fleet Replacement problems): what is the optimal interval of purchasing new vehicles,
– optimal routing (Fleet Routing problem): How should the routing of the vehicles be designed in order to minimise transport costs.
– Integrated approaches support the fleet planning process by combining more aspects of the fleet routing, replacement and composition/size problems (Mennenga 2014).

Life cycle oriented approaches are designed to assess relevant factors over the life cycle. The holistic view helps avoiding problem shifting between life phases. Life cycle oriented approaches encompass life cycle costing (LCC) and life cycle assessment (LCA) approaches. While LCC-approaches focus on the quantification of the costs and revenues during the whole life cycle, life cycle assessments consider the potential environmental impact of the fleet (Mennenga 2014; Bielli et al. 2011).

Figure 6.2 classifies approaches from the Operations Research and from Life Cycle Engineering into a structuring framework for the context of life cycle oriented fleet planning with alternatively powered vehicles. A detailed literature review of approaches from the different categories is presented in Mennenga (2014).

A number of authors examine the integration of electric vehicles into commercial vehicle fleets. Different studies have dealt with the needs of commercial fleet users and the potential benefits of electric vehicles in corporate fleets (Betz and Lienkamp 2016). Other papers deal with the arising problems when operating electric vehicles, e.g. limited range and long charging times, or specifically address the charging infrastructure on a strategic and operational level (Döppers and Iwanowski 2012). Betz and Lienkamp highlight the further need for the holistic investigation of corporate electric fleets and therefore introduce a simulation approach that supports the integration of electric vehicles into commercial fleets under consideration of the fleet-management, energy system and charging infrastructure (Betz and Lienkamp 2016).

Despite the number of studies, there is still a lack of concepts, methods and tools with a holistic view on the integration of electric vehicles into corporate fleets under consideration of the subsystems energy-, charging-, and fleet-management. Moreover, there is a further need for practical decision-support methodologies and/or guidelines that puts fleet managers in position to encounter the challenges arising from the introduction of electric vehicles to corporate fleets and make a change towards a less environmentally harmful mobility.

6.4 Decision Support Methodology for Integrating Electric Vehicles into Corporate Fleets

Decision support for integrating alternatively powered vehicles into corporate fleets requires a holistic concept which enables fleet planners to understand their needs, fleet applications, as well as possible outcomes and trade-offs of their decisions. The focus of such a concept should be on fleets that have a high potential for alternatively powered vehicles. This primarily concerns fleets in which part of their application scenarios can be covered by electric and/or plug-in hybrid vehicles and which have a fixed location for arrival and departure, i.e. that the vehicles are available at a charging point after the end of their journey. In addition, the fleet size must justify the cost of engaging decision support.

With regard to the determined requirements of fleet planners, a workshop-based concept provides decision support for integrating alternatively powered vehicles into corporate fleets (see Fig. 6.3). It consists of the four modules (1) Introduction to a holistic fleet planning, (2) Situation analysis and fleet configuration, (3) Life cycle evaluation and (4) Derivation of recommendations. The aim of the first module is to generate a mutual understanding of holistic fleet planning and the necessity of taking a life cycle perspective. The second module helps to analyse the specific situation of a fleet planer and to generate and/or prepare relevant data for a life cycle evaluation. The third module enables this evaluation by means of a simulation based decision support system for analysing and evaluating different fleet configurations under consideration of a life cycle perspective. Finally, in the fourth module the

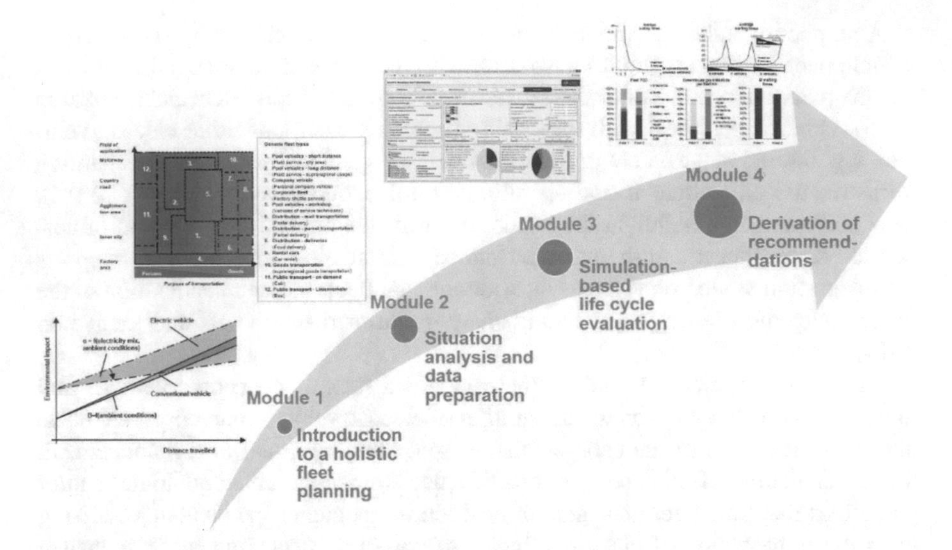

Fig. 6.3 Modules of the decision support methodology for integrating electric vehicles into corporate fleets

results of the application of the decision support systems are discussed in order to derive recommendations for fleet planners.

The workshop-based concept is the prerequisite for engaging a tool-based consulting for a specific fleet application. Thus, for each workshop module methodologies, tasks and material exist as shown in the following for each module in more detail.

6.4.1 Introduction to Holistic Fleet Planning

The aim of the first module is to provide a mutual understanding of holistic fleet planning and the necessity to take a life cycle perspective. It serves the development of a common knowledge base for all people who participate in the fleet planning process. The fleet planning process aims to identify, analyse and evaluate different alternatives for the composition of the fleet vehicle stock and the organization of the fleet vehicle operation so that the vehicle fleet fulfils functional, economic and environmental viability. In order to address the aforementioned objective, it is essential that fleet planners are aware of the necessity to take a life cycle perspective. Figure 6.4 indicates the need for taking a life cycle perspective and it highlights the interdependencies between the life phases and the objective dimensions.

The fleet life cycle divides into three main phases: fleet production including raw material extraction and manufacturing, fleet usage and fleet end-of-life or recycling. As part of the fleet usage, the fuel and/or electricity supply chain must be taken into

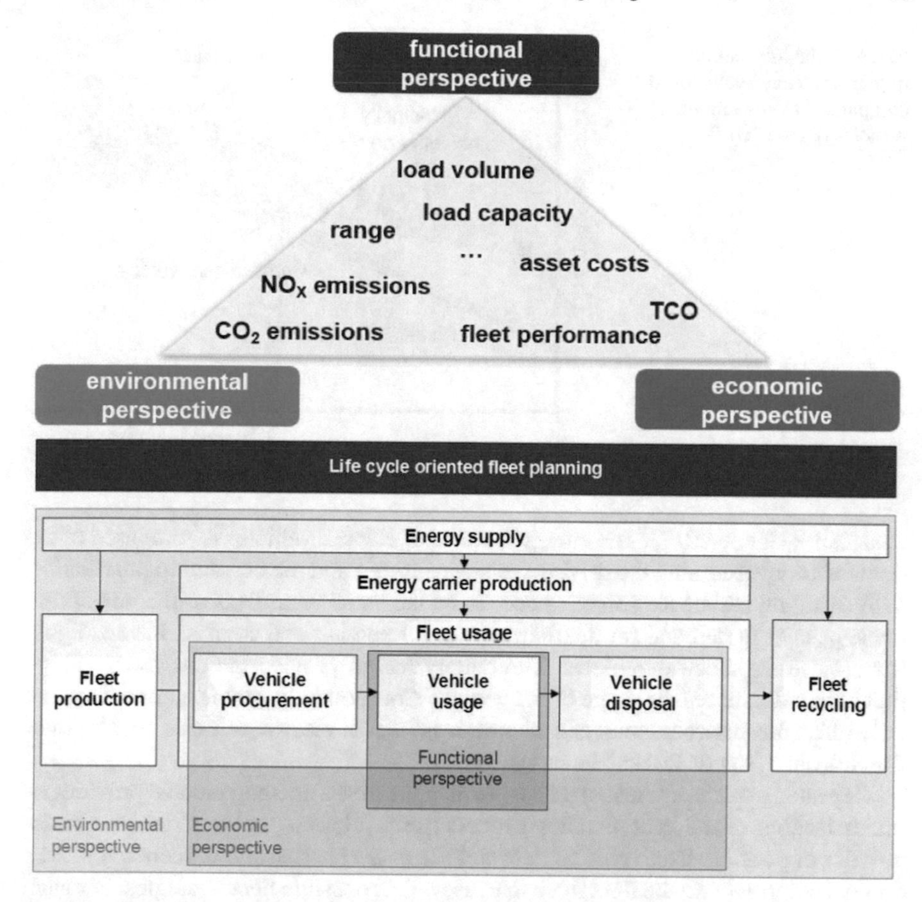

Fig. 6.4 Life cycle perspectives of a vehicle fleet (Mennenga 2014)

account as well so that a complete assessment of the impacts of fleet usage is possible (Mennenga 2014).

Phase-oriented life cycle concepts support the visualization of the chronological correlation of processes, the identification of goal-conflicts over the life cycle as well as the identification of relevant influencing factors (Herrmann 2010). Exemplarily, Fig. 6.5 gives a qualitative representation of the environmental impact of an electric and a conventional vehicle. It reflects possible environmental advantages and disadvantages of electric vehicles. Firstly, the production of electric vehicles potentially has a higher environmental impact than the production of conventional vehicles. This mainly results from the battery production with a share of around 35–41% (Cerdas et al. 2016; Ellingsen et al. 2014, 2016). Secondly, the environmental impact of the use phase highly depends on the electricity mix and the energy demand of the vehicle (Egede 2017). Thus, also its environmental impact highly depends on the source of energy that is used to power the vehicles (renewable or non-renewable) (Nordelöf

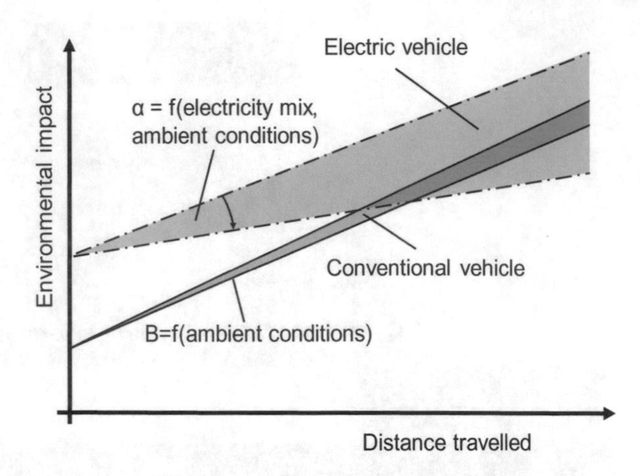

Fig. 6.5 Environmental impact of electric vehicles in comparison to conventional vehicles (Egede 2017)

et al. 2014; Egede 2017). The energy demand of electric vehicles is composed of the demand for overcoming the driving resistance forces and the demand for auxiliaries. While driving resistance forces appear to be the same regardless of the powertrain concept, energy demand for auxiliaries, e.g. for heating and cooling, have a higher importance for electric vehicles. The battery has to provide the necessary energy that in turn decreases the range of the vehicle. Consequently, ambient conditions not only affect the environmental and economic impact of electric vehicles but also their functionality (Egede 2017; Mennenga 2014).

Depending on the interest of a fleet planner, the first workshop module provides an understanding of the fleet planning process itself, relevant goal criteria and characteristics of possible fleet vehicles, relevant life cycle phases, and influencing factors on goal criteria along the fleet life cycle, as well as possible fleet strategies. For each topic, elaborated workshop methodologies support the transfer of knowledge, e.g. thought-provoking presentations or group brainstorming. They help to impart when electric vehicles could be a better choice from a functional, economic and environmental perspective and to show the variety of influencing factors on this decision.

6.4.2 Situation Analysis and Data Preparation

The second workshop module helps to analyse the specific situation of a fleet planner and to generate and/or prepare relevant data for a life cycle evaluation. Companies, especially SMEs, usually do not have a comprehensive overview of their fleets prior to the beginning of the situation analysis. Therefore, the objective of the second workshop module is to collect and structure all relevant data. The information serves as a necessary input for the following simulation (see Sect. 6.4.3) as well as for the evaluation of the simulation results (see Sect. 6.4.4). The data acquisition covers the areas of fleet planning with regard to general constraints and circumstances of the fleet composition and operation.

Relevant data about the general constraints and circumstances of fleet planning include but are not limited to charging infrastructure, existing contracts and their duration with car dealers, budget and corporate strategy. Data about the fleet composition and operation can help better understand the actual needs of the company and discover potential improvements. The data can be broken down into functional, economic and environmental categories. To name a few examples, frequency of usage, distances travelled, required seats, load volume and weight belong to functional criteria. Data about the economics of the vehicle fleet covers the direct and indirect costs of operation. The ecological performance of the fleet is typically measured by the fleet average CO_2 emissions. In the case of electric vehicles, the electricity mix allows for calculating the indirect emissions and is therefore a relevant factor.

The workshop focuses not only on the identification of relevant data but also on the definition of the data acquisition process itself, as well. Although most of the required data is already present in the company, the definition of the data acquisition process is necessary, as the data is split between different departments and it is usually not known, where to find the data.

In order to support the data acquisition process, an excel-based data acquisition sheet has been developed that provides an understanding of what data is needed in which quality. The sheet provides a framework for assessing relevant data ranging from the boundaries of fleet planning over the fleet structure to fleet operation. It also covers the topics of searching for vehicle alternatives and goals as well as evaluation criteria. Here, the quality of the available data is crucial. Therefore, a step within the workshop focuses on checking for controversial facts and filling out white spots. After the collection of the necessary data, it has to be summarized and presented in a form that allows for further processing the data to the simulation. Here the excel sheet also provides structural support.

The output of the second workshop module is a fundamental understanding about current fleet operation practices of the company, as well as circumstances and constraints. The structured process allows for the identification of first improvement measures as well as unfolding possibilities for integrating alternatively powered vehicles into the fleet. These measures and alternative fleet sizes and/or compositions can be simulated and evaluated in the next step.

6.4.3 Simulation Based Life Cycle Evaluation

The third part of the workshop-based concept refers to the application of an existing simulation based decision support system for life cycle oriented vehicle fleet planning (Mennenga 2014). Methodologically, the decision support system integrates a form-based process guide, an agent-based simulation of the fleet operation as well as the life cycle evaluation methodologies demand-driven life cycle costing or total cost of ownership and life cycle assessment. The analytical models for TCO and LCA

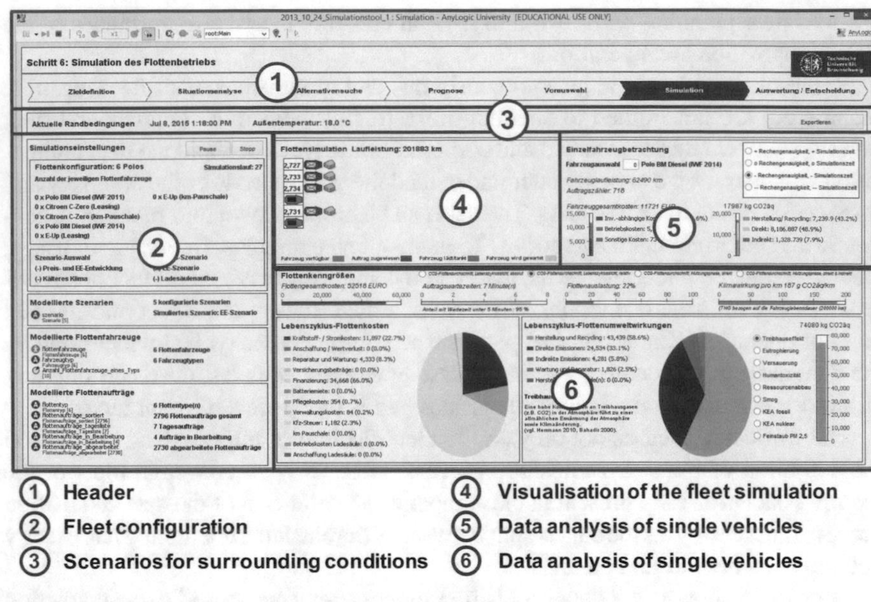

Fig. 6.6 Simulation screen of the decision support system (Mennenga 2014)

and the agent-based model are embedded in a suitable analysis and visualisation environment in MS Excel and Anylogic software.

The decision support system is designed as an expert system and thus is not designed to be used by a fleet planer of a SME. Instead, a fleet consultant models and evaluates different fleet configurations of potential vehicle fleets with the help of the gathered data and information from Sect. 4.2. To do so, the consultant incorporates data for the fleet strategy, the fleet application, possible fleet vehicles as well as surrounding conditions of the fleet. With this data, it is possible to configure different fleet configurations and simulate the fleet operation under different surrounding conditions. Exemplarily Fig. 6.6 shows the simulation screen of the decision support system. It provides information regarding the operation of the vehicle fleet, key figures for single vehicles and the vehicle fleet. This comprises functional, economic and environmental criteria and reflects a life cycle perspective. After one simulation run is finished all data for the specific fleet configuration and the underlying operating conditions are stored in a database. In this way, different fleet configurations may be compared to each other regarding their fulfilment of goal criteria.

For more details regarding the simulation please refer to Mennenga (2014) and Mennenga and Herrmann (2016).

6.4.4 Derivation and Transfer of Recommendations

The fourth workshop module serves the presentation, comparison, evaluation and discussion of the simulation results of different simulation runs. This enables the derivation and transfer of recommendations regarding the composition of a specific vehicle fleet and the organization of the fleet vehicle deployment regarding the fulfilment of goal criteria.

Different evaluation options allow for imparting knowledge from the simulation results to fleet planners. In Fig. 6.7 four kinds of evaluation options are illustrated with exemplary evaluation sheets (Mennenga 2014):

The first evaluation option refers to the determination of the required fleet stock and composition, i.e. the quantity of required vehicles and the distribution of different vehicle concepts. To this end, fleet waiting times can be analysed. They refer to the time a driver has to wait for a vehicle because there is no vehicle available, at

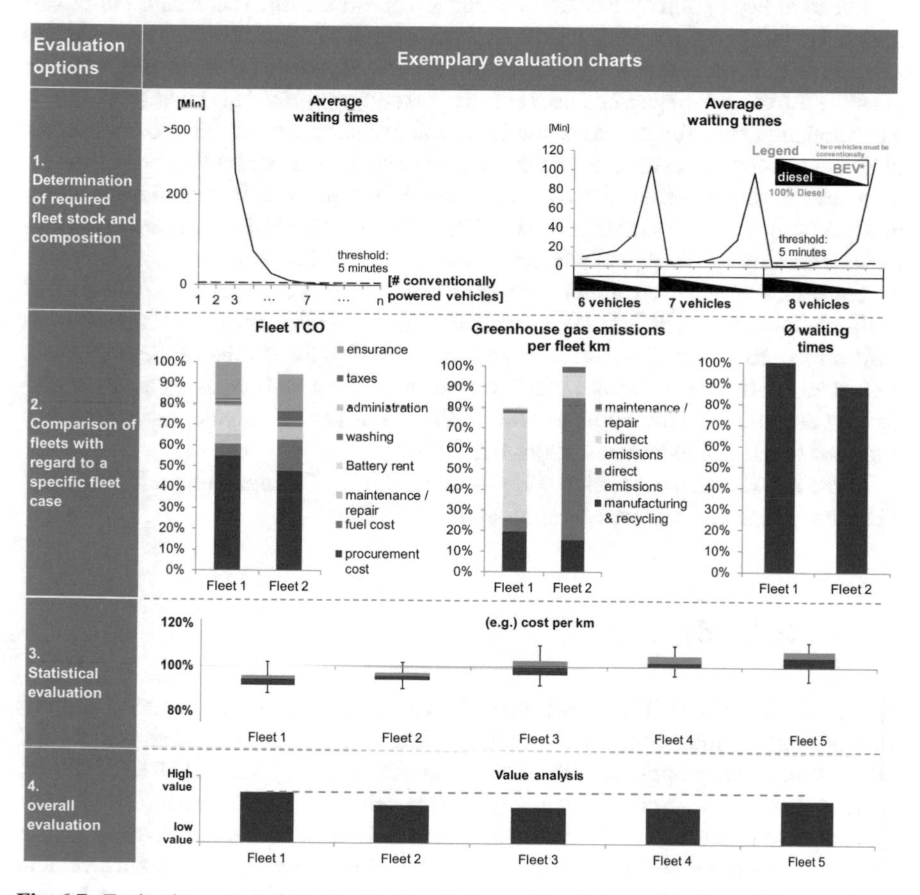

Fig. 6.7 Evaluation options for imparting knowledge to fleet planners, based on Mennenga (2014)

least no vehicle with enough range. The waiting times are analysed for a fleet of solely conventionally powered vehicles. In order to estimate the required minimum vehicle stock the number of conventionally powered vehicles is increased with every simulation run. Afterwards, conventionally powered vehicles are replaced by other vehicle concepts such as battery electric vehicles. This enables the evaluation of their influence on the fleet operation. Here, the influence of charging times becomes obvious.

The second evaluation is the comparison of fleets with regard to a specific fleet case. Different evaluation criteria, such as waiting times, life cycle costs, or life cycle emissions can be used. In this way, it is possible to evaluate whether a fleet performs better than a reference fleet under consideration of one specific fleet application and scenario. Furthermore, the distribution of life cycle costs and emissions can be analysed with regard to the different cost components or life cycle phases. This allows for identifying potential trade-offs between cost factors or problem shifting from one life cycle phase to another.

The third evaluation option refers to statistical evaluation. The evaluation of various fleet configurations is subject to uncertainties. Consequently, they are modeled in a stochastic simulation model. Thus, the results of a simulation run are only one possible result for the specified fleet application and scenario. This means that different simulation runs for the same configuration and scenario can lead to completely different simulation results. In particular, if the results of different fleet scenarios are compared it must be ensured that different results are statistically significant and not just a consequence of stochastic effects. Thus, several simulation runs are necessary for determining the fleet performance of one single fleet configuration.

The fourth evaluation option finally gives the overall evaluation. The overall evaluation gives insight into the value proposition of different fleet configurations with respect to a specified fleet application. Typically, the results of the previously described evaluation options influence this overall evaluation. In general, multi-criteria evaluation methodologies can be used. Further information is included in e.g. (Götze 2014; Geldermann 2006; Riedl 2006).

In the following, the decision support methodology for integrating electric vehicles into corporate fleets will be applied to a case study.

6.5 Case Study

As part of the research project Fleets Go Green, the presented decision support methodology for integrating electric vehicles into corporate fleets has been developed and simultaneously applied to the fleet of the local energy supplier BS|ENERGY. As it is typical for a service company with a vehicle fleet, the reduction of costs and a higher utilization of the vehicles was highly relevant for BS|ENERGY. The application of the developed methodology aimed at analysing an existing vehicle fleet and evaluating whether electric vehicles could replace conventional vehicles.

As the workshop methodology has been developed over the course of Fleets Go Green, the concerns and needs of the company were assessed in several project meetings. In order to bring the participants to a common understanding and raise awareness for the influencing factors, the essential information about fleet planning and life cycle thinking was shared in these meetings via thought-provoking presentations and discussed afterwards within the project team.

The aim of the situation analysis and data generation was to find out how the fleet is used and to prepare the data for the subsequent simulation. In total, the fleet of the company consisted of over 150 vehicles. In order to decrease complexity, the case study focused on an extract of this fleet. This extract consisted of 22 vehicles that are located at a location separate from the rest of the fleet. From the 22 vehicles, fifteen vehicles were directly assigned to a specific driver. The remaining seven vehicles create a pool that was accessible for different departments for the use in daily business. The fleet in its current state consists of four Volkswagen Caddy and three Volkswagen Polo that are all conventionally powered. The basis for the data acquisition was the fleet that solely consisted of conventional vehicles, as this was the starting point for the planning process. The availability, quality and resolution of data strongly influences the outcome of the fleet usage assessment. Here, the excel-based data acquisition sheet was used to assist in this process. The data acquisition was done based on manually documented logbooks. Seven of them were used exemplarily for review. They were transferred to Excel and analysed regarding the fleet usage afterwards. As the logbooks did not consequently contain sufficient information about the trip destinations, the fleet tasks were categorized into two fleet types (1) daily service and (2) on-call service.

Figures 6.8 and 6.9 show the results from the analysis of the logbooks. Figure 6.8 depicts the cumulated share of trips in comparison to the distance travelled. The diagram distinguishes between daily business-trips and on-call service-trips. It shows that about 60% of the total number of trips have distances of less than 15 km and 90% of the trips are within a distance of 50 km. Moreover, daily business trips only last for 12 km on average. On-call service trips have an average distance of 52 km. However, on-call service trips have not been properly documented in the logbooks because it was not apparent how many single trips were taken together and documented as one on-call service trip.

Figure 6.9 outlines the frequency of fleet usage over the period of a day. To this end, the frequency of trips was calculated over a day for each 15-min period. The diagram distinguishes between daily business and on-call service-trips. The distribution of the daily business trips shows a peak period at the beginning of the day shift. Afterwards, the frequency of usage is continuously falling down until midday. Thereafter, it stays on a nearly constant level until the end of the day shift. The distribution shows a second peak shortly before midnight. Afterwards, the usage of the vehicles falls close to zero. The distribution of the on-call service-trips has on the contrary a high peak at the end of the day shift. The rest of the day, it stays close to zero. One explanation for the peak could be that the on-call duty employees take the vehicles home.

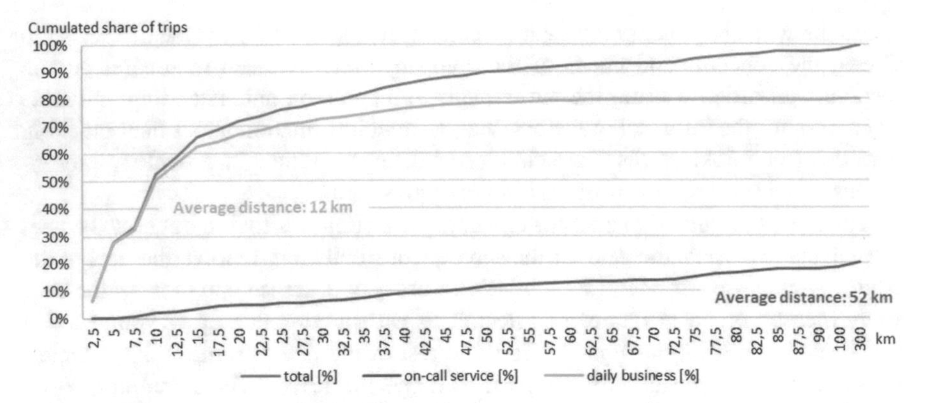

Fig. 6.8 Cumulated share of trips in the analysed fleet

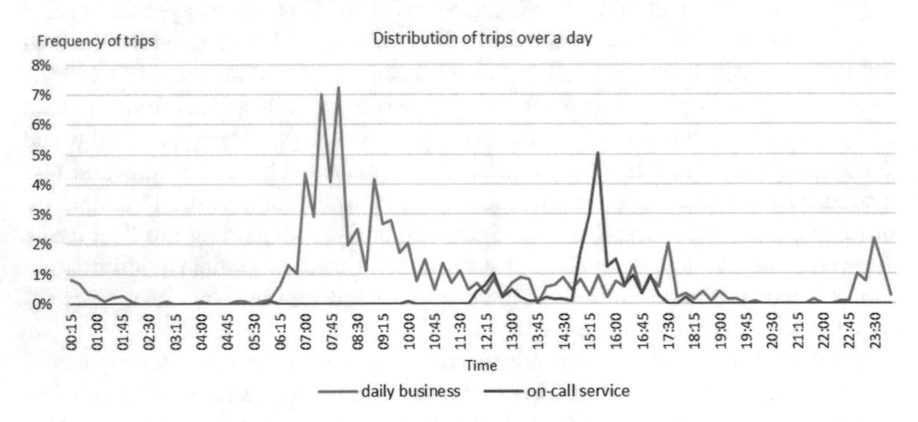

Fig. 6.9 Distribution of trips over a day

The next step was the conduction of the simulation based life cycle evaluation. To this end, the seven vehicles that create a pool were considered as the others were directly assigned to a specific driver and this assignment was not part of the evaluation. First, the required fleet stock and composition of the fleet pool has been determined. To this end, several simulation runs have been carried out, each with a different fleet size and composition. First, fleet waiting times have been analysed. Waiting times below five minutes were defined as the required target for the functionality of the fleet. The simulation underlay the following assumptions: The time horizon of the simulation was set to one year. The lifetime of vehicles has been 200,000 km, afterwards, they would be replaced by new ones. The location of the fleet has already been equipped with charging stations. Up to five electric vehicles could be charged at the same time. Further investments in charging infrastructure were not planned within then planning horizon. Based on these assumptions Fig. 6.10 depicts the fleet waiting times and the corresponding fleet size. For reaching the threshold of a five-minute waiting interval, at least five vehicles are required. Thus, it seems that the pool

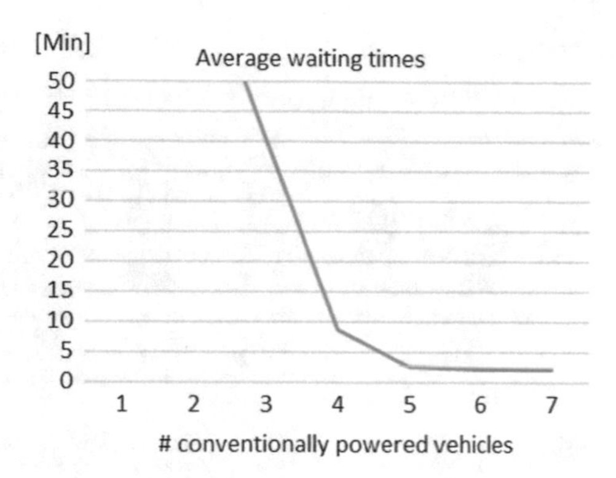

Fig. 6.10 Determination of fleet size for conventionally powered vehicles

fleet is oversized with seven vehicles. However, the case study does not fully address the question of the carried load in vehicles. Due to information gaps, distinguishing between uses cases with different loads was not possible. Therefore, carried load should be part of the analysis for obtaining more precise results, as well.

In the next step, conventionally powered vehicles were replaced by electric vehicles. Figure 6.10 outlines only small differences in average waiting times between fleet sizes of five, six and seven vehicles. Therefore, two fleets consisting of five and six conventionally powered vehicles served as the basis for implementing electric vehicles. Conventionally powered vehicles were replaced one by one with electric vehicles. Additionally, a fleet consisting of only four electric vehicles was analysed, as well. Figure 6.11 shows the resulting fleet waiting times and the corresponding fleet sizes and compositions. As expected, fleet waiting times increase with the integration of electric vehicles, e.g. due to charging times. However, Fig. 6.11 also shows fleet configurations with a higher share of electric vehicles and in total less vehicles that seem to be more advantageous than the current fleet. However, the differences in the average waiting times between the alternatives are altogether rather minor. It can be seen on Fig. 6.11 that increasing the number of Volkswagen E-Up!'s and consequently decreasing the number of Renault Kangoo Z.E.'s in an fully electric fleet results in shrinking waiting times. One explanation for this phenomenon could be the higher energy consumption of a Renault Kangoo Z.E., which in turn increases downtimes due to charging. Therefore, four of the aforementioned alternatives, that seem to be more advantageous, were chosen for further analysis. They are highlighted in Fig. 6.11, as well.

Table 6.1 lists detailed information of the chosen alternatives and the current fleet. As already mentioned before, the current fleet consists of seven conventionally powered vehicles. The first alternative configuration has both conventionally powered and electric vehicles. The total number of vehicles is however reduced by one in comparison to the current fleet. The other chosen alternatives only consist of electric vehicles. Each of them have a Renault Kangoo Z.E. and a shrinking number of Volkswagen E-Up!'s. Here, the number of E-Up!'s is reduced from five in Alternative 2 to three in Alternative 4. The planning horizon has been adapted and increased

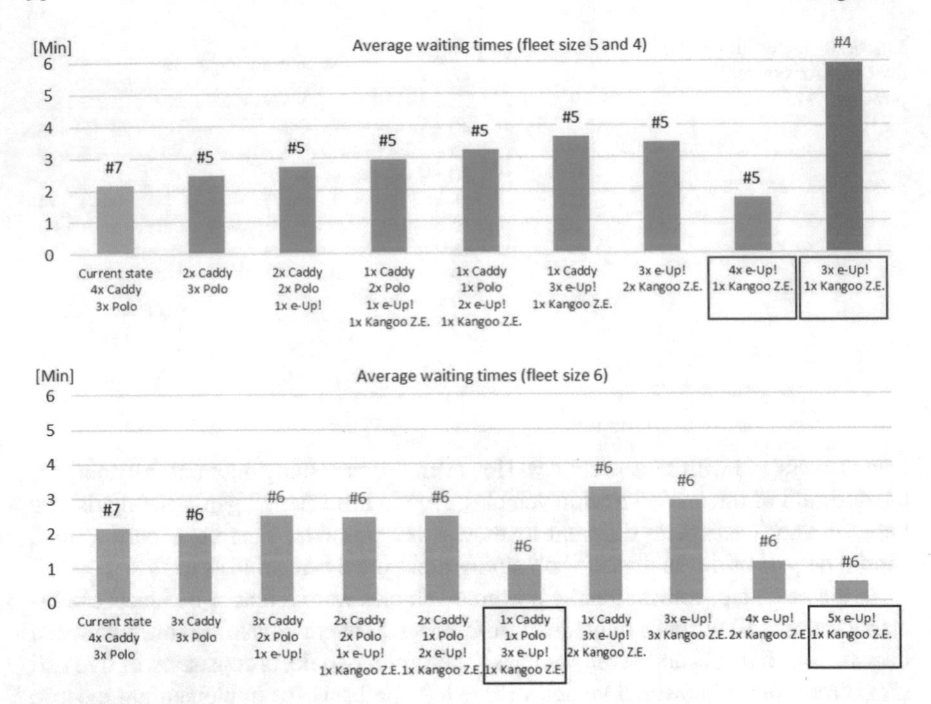

Fig. 6.11 Determination of fleet size and composition for a mixed/electric vehicle fleet

from one year to three years for the following simulation that allows for in depth comparison of the alternative fleets. The calculation of greenhouse gas emissions for electric vehicles are based on the German electricity mix from the year 2012. Energy demands for heating and cooling were reckoned based on Central German climate. All other simulations constraints remained the same.

Figure 6.12 provides an overview of the simulation results. It shows a comparison of the current fleet with the four aforementioned alternative fleets from a functional, ecological and economic perspective. As expected, the fleet TCO decrease with a declining number of vehicles while still maintaining the functionality of the fleet. Leasing costs are the most dominant cost factor. Therefore, decreasing the number of vehicles has a high contribution to lowering cost over the life cycle. On the contrary, the implementation of electric vehicles has a negative effect to the TCO due to higher leasing rates. In the case of the simulated alternative fleet configurations the negative effect of higher leasing rates could however be evened out by decreasing the fleet stock. Each fleet configuration expect Alternative 4 has lower waiting times on average than the target of five minutes. While Alternative 4 exceeds the target of five minutes, it has on the contrary the lowest TCO and its environmental performance is the best in comparison to other alternatives. Therefore, Alternative 4 could still be a suitable choice depending on the objectives of the fleet management of BS|ENERGY. The low TCO and greenhouse gas emissions on the one hand and the high waiting times on the other hand result from the smaller fleet size of Alternative 4 with only

Table 6.1 Current and alternative fleet configurations

					# electric vehicles	# total
Current fleet	4 VW Caddy CNG	3 VW Polo			0	7
Alternative 1	1 VW Caddy CNG	1 VW Polo	3 VW E-Up!	1 Renault Kangoo Z.E.	4	6
Alternative 2			5 VW E-Up!	1 Renault Kangoo Z.E.	6	6
Alternative 3			4 VW E-Up!	1 Renault Kangoo Z.E.	5	5
Alternative 4			3 VW E-Up!	1 Renault Kangoo Z.E.	4	4

four vehicles. Figure 6.12 depicts further the relation of greenhouse gas emissions over the life cycle for the simulated fleets. The results indicate again that the current fleet has improvement potential. All analysed alternatives have better environmental performance than the current fleet. Similar to the fleet TCO, it can be seen that the substitution of a conventionally powered vehicle by an electric vehicle results in an increase in manufacturing related emissions while decreasing direct emissions. The depicted improvement of the alternative fleet configurations is levered by a combination of a decrease in direct emissions, an increase in manufacturing related emissions and subsequent decrease in fleet stock which in turn results in a decrease in total.

Figure 6.13 shows the result of an overall evaluation of the simulated fleets. The value of each configuration was calculated by comparing fleet waiting times, TCO and greenhouse gas emissions per fleet km. Each of them have been weighted equally. Figure 6.13 indicates again that the current fleet has improvement potential as the alternative fleet configurations scored better in the simulation.

Summing up the results, Alternative fleet 2 and 3 seem to be good choices and have been therefore analyzed statistically. Statistical evaluation is required to address uncertainties regarding functional, economic and ecological performance indicators in the results. Figure 6.14 shows the corresponding results of the statistical evaluation. As expected, Alternative 3 has a lower TCO, as this fleet consists of less vehicles. As a consequence of it, average waiting times are higher than those of Alternative 2. However, the higher waiting times are still not critical for the availability of the fleet, as they are still within the target of five minutes. Due to the decreased number of vehicles in Alternative 3, greenhouse gas emissions are as well lower than in Alternative 2. Although Alternative 3 performs overall better than Alternative 2, the insecurities regarding the results are higher. Higher insecurities regarding waiting

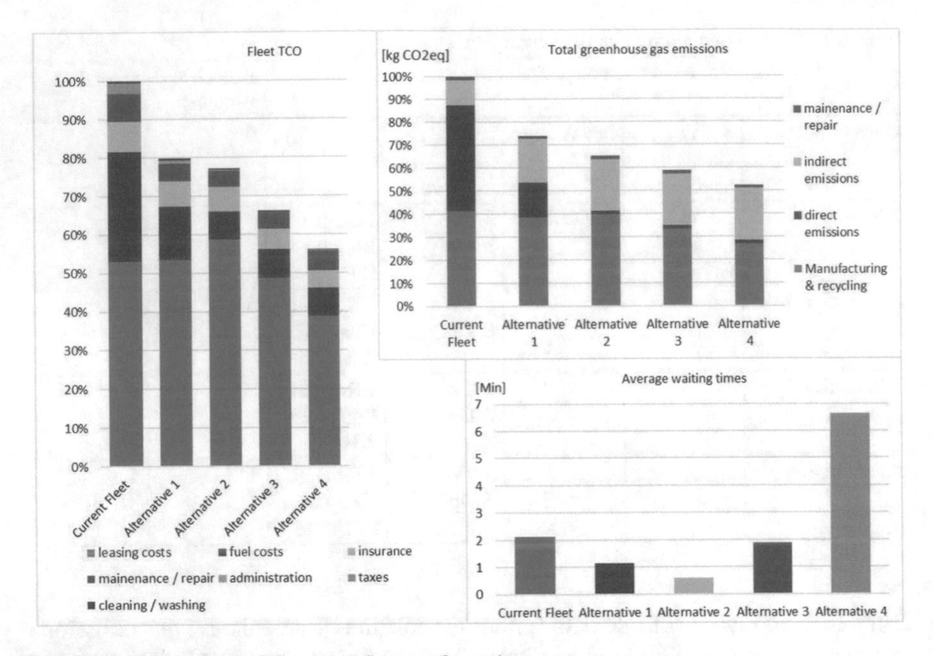

Fig. 6.12 Comparison of simulated fleet configurations

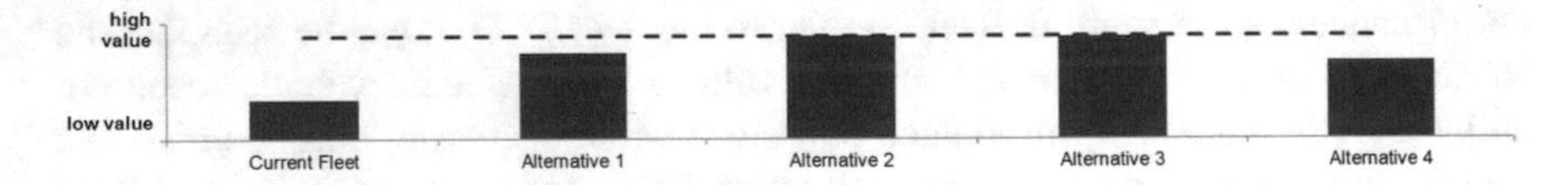

Fig. 6.13 Overall evaluation of the examined fleets

times are due to the smaller vehicle stock in Alternative 3. Depending on the amount and timing of fleet jobs, the smaller vehicle stock may be overloaded and therefore fleet jobs may be postponed. Furthermore, both, TCO and greenhouse gas emissions indicate that the insecurities regarding fleet performance increase with decreasing fleet stock.

The results indicate that electric vehicles are suitable for the fleet under the analysed circumstances. Moreover, it has been shown that the current fleet consisting of seven vehicles seem to be oversized for its task. Therefore, decreasing the number of vehicles and making a transition towards electric vehicles may be a viable solution to improving the fleet performance. As it has been mentioned before, the case study does not fully address the question of the carried load in vehicles. Due to information gaps, distinguishing between uses cases with different loads was not possible. In order to adequately address this problem, further fleet types (1a) daily service with higher carried load (1b) daily service with lower carried weight have to be introduced and simulated analogously.

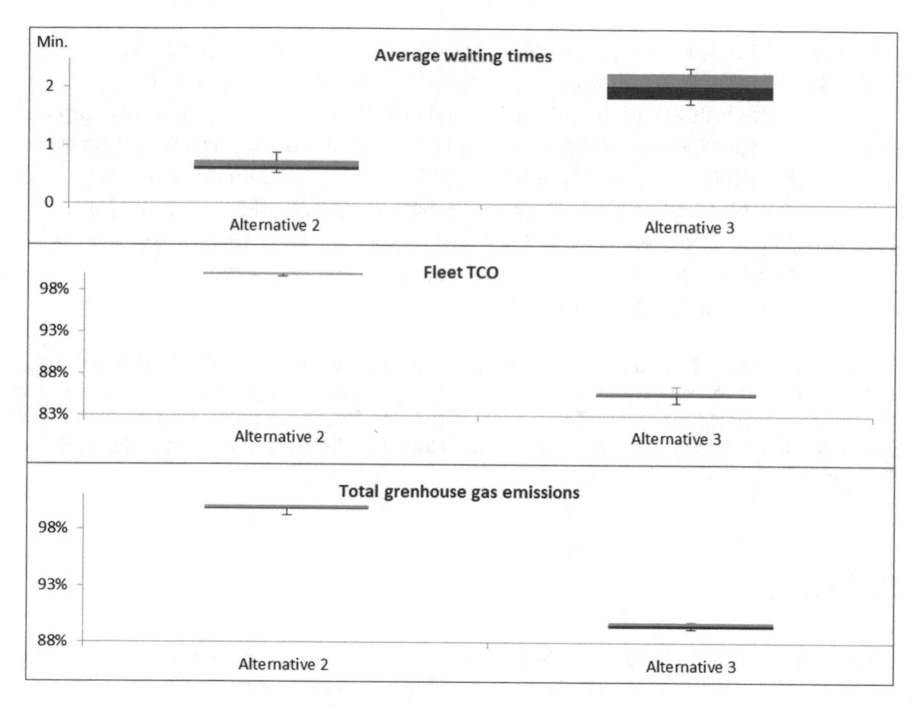

Fig. 6.14 Statistical evaluation of fleet alternative 2 und 3

6.6 Summary

A workshop based decision support methodology for the integration of alternatively powered vehicles into corporate fleets is introduced. The analysis of decision-making in corporate fleet planning and a literature review are the basis for illustrating the proposed concept. The concept takes into account the fleet size and fleet operation planning as well as a lifecycle perspective. It is based on a workshop methodology that consists of four modules: introduction to holistic fleet planning, situation analysis and fleet configuration, life cycle evaluation and derivation of recommendations. The modules reflect necessary steps of a lifecycle oriented fleet planning and help the fleet planner to investigate necessary data and evaluate different fleet configurations based on a fleet simulation. Finally, recommendations can be given for the composition and operation of the fleet in general and in particular for the integration of electric vehicles into a current fleet. The methodology combines the functional, economic and environmental perspective in the fleet planning with alternatively powered vehicles. It specifically supports decision making for fleets of small and medium size enterprises (e.g. craft fleets, delivery services, car-sharing providers). The methodology is presented within a case study for the fleet of a local energy supplier. Data acquisition proved to be the most critical factor during implementation. In order to support a valid fleet planning and to prevent data gaps resulting from insufficiently documented logbooks of vehicles, continuous measurement of fleets is recommended.

For the further development of the presented methodology, the combined consideration of the subsystems energy, charging and fleet planning with a special scope on police applications is subject of the project lautlos&einsatzbereit (silent&ready4use) supported by the German Federal Ministry for the Environment, Nature Conservation, Building and Nuclear Safety. The project objective is to develop guidelines for the economic and ecological planning, controlling and operation of mixed vehicle fleets in highly demanding situations.

Acknowledgements The authors express their gratitude to the German Federal Ministry for the Environment, Nature Conservation, Building and Nuclear Safety for supporting the project "Fleets Go Green—Integrated analysis and evaluation of the environmental performance of electric and plugin-hybrid vehicles in everyday usage on the example of fleet operations" under the reference 16EM1041.

References

Betz J, Lienkamp M (2016) Approach for the development of a method for the integration of battery electric vehicles in commercial companies, including intelligent management systems. Automot Engine Technol 1(1–4):107–117. https://doi.org/10.1007/s41104-016-0008-y

Bielli M, Bielli A, Rossi R (2011) Trends in models and algorithms for fleet management. Procedia-Soc Behav Sci 20:4–18. https://doi.org/10.1016/j.sbspro.2011.08.004

Cerdas F, Mennenga M, Herrmann C (2016) Model-based life cycle engineering framework for electric vehicles In: 4th joint-symposium, Advanced Vehicle Energy Concepts and Structures for China (AVECS). Braunschweig, Institut für Konstruktionstechnik, Braunschweig, pp 82–91. ISBN 978-3-9816886-9-6

Döppers F-A, Iwanowski S (2012) E-mobility fleet management using ant algorithms. Procedia-Soc Behav Sci 54:1058–1067. https://doi.org/10.1016/j.sbspro.2012.09.821

Egede P (2017) Environmental assessment of lightweight electric vehicles. Springer International Publishing, Cham

Egede P, Dettmer T, Herrmann C, Kara S (2015) Life cycle assessment of electric vehicles—a framework to consider influencing factors. Procedia CIRP 29:233–238. https://doi.org/10.1016/j.procir.2015.02.185

Ellingsen LA-W, Majeau-Bettez G, Singh B, Srivastava AK, Valøen LO, Strømman AH (2014) Life cycle assessment of a lithium-ion battery vehicle pack. J Ind Ecol 18(1):113–124. https://doi.org/10.1111/jiec.12072

Ellingsen LA-W, Singh B, Strømman AH (2016) The size and range effect: lifecycle greenhouse gas emissions of electric vehicles. Environ Res Lett 11(5):54010. https://doi.org/10.1088/1748-9326/11/5/054010

Geldermann J (2006) Mehrzielentscheidungen in der industriellen Produktion. KIT Scientific Publishing

Gnann T, Plötz P, Zischler F, Wietschel M (2012) Elektromobilität im Personenwirtschaftsverkehr – eine Potenzialanalyse: Working Paper Sustainability and Innovation No. S 7/2012. http://www.isi.fraunhofer.de/isi-wAssets/docs/e-x/de/working-papers-sustainability-and-innovation/WP07-2012_Wirtschaftsverkehr.pdf. Accessed 8 Mar 2017

Götze U (2014) Investitionsrechnung. Springer, Berlin

Herrmann C (2010) Ganzheitliches Life Cycle Management: Nachhaltigkeit und Lebenszyklusorientierung in Unternehmen. VDI-Buch. Springer, Berlin

IEA (2016) Global EV Outlook 2016: beyond one million electric cars. https://www.iea.org/publications/freepublications/publication/Global_EV_Outlook_2016.pdf. Accessed 8 Mar 2017

Kaiser R (2014) Qualitative Experteninterviews: Konzeptionelle Grundlagen und praktische Durchführung. Lehrbuch. Springer, Wiesbaden

Kasten P, Zimmer W, Leppler S (2011) CO2-Minderungspotenziale durch den Einsatz von elektrischen Fahrzeugen in Dienstwagenflotten: Ergebnisbericht im Rahmen des Projektes "Future Fleet" AP 2.7. http://www.oeko.de/oekodoc/1343/2011-027-de.pdf. Accessed 8 Mar 2017

Kerler S (2003) Fuhrpark- und Flottenmanagement: Praxishandbuch für Unternehmen mit eigenem Fuhrpark. Vogel, München

Kraftfahrt-Bundesamt (2015) Neuzulassungen von Pkw in den Jahren 2006 bis 2015 nach ausgewählten Haltergruppen. http://www.kba.de/DE/Statistik/Fahrzeuge/Neuzulassungen/Halter/z_n_halter.html?nn=652344

Kreyenberg D (2016) Fahrzeugantriebe für die Elektromobilität. Dissertation, Springer Fachmedien Wiesbaden GmbH

LeaseTrend AG (2012) C A T I - S T U D I E unter 100 Flottenmanagern zum Thema Flotte der Zukunft. http://www.autoleasing.de/download/LeaseTrend_CATI_Flotte_der_Zukunft_0812.pdf. Accessed 8 Mar 2017

Mennenga MS (2014) Lebenszyklusorientierte Flottenplanung mit alternativ angetriebenen Fahrzeugkonzepten. Zugl.: Braunschweig, Techn. Univ., Diss., 2014. Schriftenreihe des Instituts für Werkzeugmaschinen und Fertigungstechnik der Technischen Universität Braunschweig. Vulkan-Verl., Essen

Mennenga M, Herrmann C (2016) Life cycle oriented fleet planning with alternatively powered vehicles. In: 3rd joint-symposium, Advanced Vehicle Energy Concepts and Structures for China (AVECS)

Nesbitt K, Sperling D (2001) Fleet purchase behavior: decision processes and implications for new vehicle technologies and fuels. Transp Res Part C: Emerg Technol 9(5):297–318. https://doi.org/10.1016/S0968-090X(00)00035-8

Nordelöf A, Messagie M, Tillman A-M, Ljunggren Söderman M, van Mierlo J (2014) Environmental impacts of hybrid, plug-in hybrid, and battery electric vehicles—what can we learn from life cycle assessment? Int J Life Cycle Assess 19(11):1866–1890. https://doi.org/10.1007/s11367-014-0788-0

Riedl R (2006) Analytischer Hierarchieprozess vs. Nutzwertanalyse: Eine vergleichende Gegenüberstellung zweier multiattributiver Auswahlverfahren am Beispiel Application Service Providing

Schrempf EC (2015) Fuhrparkmanagement – ein ökonomisches und ökologisches Konzept. In: Stenner F (ed) Handbuch Automobilbanken. Springer, Berlin, pp 227–238

Statista (2016a) Anzahl der Fahrzeuge im Fuhrpark von Unternehmen in Deutschland nach Unternehmensgröße. https://de.statista.com/statistik/daten/studie/476870/umfrage/unternehmen-anzahl-der-fahrzeuge-im-fuhrpark/. Accessed 8 Mar 2017

Statista (2016b) Verteilung der Neuzulassungen von Personenkraftwagen in Deutschland im Jahr 2015 nach gewerblichen und privaten Haltern. https://de.statista.com/statistik/daten/studie/5098/umfrage/anteil-von-privaten-gewerblichen-haltern-bei-pkw-neuzulassungen/. Accessed 8 Mar 2017

VMF (2017) Der deutsche Flottenmarkt. http://www.vmf-fuhrparkmanagement.de/de/Branche-und-Markt/Flottenmarkt-Deutschland. Accessed 8 Mar 2017

Vogel M (2016) Elektromobilität in gewerblichen Anwendungen. https://difu.de/node/9580. Accessed 5 Jan 2017

Chapter 7
Recommendations from Fleets Go Green

Christoph Herrmann, Michael Bodmann, Stefan Böhme, Antal Dér, Selin Erkisi-Arici, Ferit Küçükay, Michael Kurrat, Daniela Mau, Mark Stephan Mennenga, Jan Mummel, Marcel Sander and David Woisetschläger

7.1 Introduction

Electric vehicles have the potential to reduce emissions from road transport, while releasing no local emissions during the use phase. The utilization of electric vehicles in fleet operations offers an excellent opportunity for the rapid diffusion of electric vehicles into the market due to the fast turnover rate of fleet vehicles. However, further research is necessary to examine the utilization of electric vehicles in daily use in order to recognize drawbacks and to determine further improvement potentials. The project Fleets Go Green aims to study the environmental assessment of electric vehicles in fleet operations. Fleets Go Green consists of different research modules,

---

C. Herrmann · S. Böhme · A. Dér · S. Erkisi-Arici · M. S. Mennenga (✉)
Institute of Machine Tools and Production Technology (IWF), Technische Universität Braunschweig, Braunschweig, Germany
e-mail: m.mennenga@tu-braunschweig.de

M. Bodmann
TLK-Thermo GmbH, Braunschweig, Germany

F. Küçükay
Institute of Automotive Engineering, Technische Universität Braunschweig, Braunschweig, Germany

M. Kurrat · J. Mummel
Institute of High Voltage Technology and Electrical Power Systems, Technische Universität Braunschweig, Braunschweig, Germany

M. Sander
Technische Universität Braunschweig, Institute of Automotive Engineering (IAE), Braunschweig, Germany

D. Mau · D. Woisetschläger
Institute of Automotive Management and Industrial Production, Technische Universität Braunschweig, Braunschweig, Germany

C. Herrmann et al. (eds.), *Fleets Go Green*, Sustainable Production, Life Cycle Engineering and Management,
https://doi.org/10.1007/978-3-319-72724-0_7

which investigate the integrated vehicle, usage, and power supply system behavior. The total energy requirements of fleet vehicle operations with different topologies over the use phase are determined in Module 1, while the user acceptance both from fleet owners and from drivers perspectives are researched in Module 2. Module 3 aims to integrate the electric vehicle fleets in the electrical distribution system and maximize the integration of renewable energy sources in their supply. The environmental assessment of fleets is studied in Module 4. Furthermore, all findings are integrated into a decision support system for the ecologically oriented fleet management and planning in Module 5.

This paper represents the main recommendations, which could be derived from the results of the project Fleets Go Green. The recommendations target fleet owners, vehicle manufacturers and political decision makers. They underline the important findings in order to overcome the obstacles in the adoption of electric vehicle fleets, to correct potential wrong judgements, to enlighten the ecological aspects and therefore to encourage their acceptance.

7.2 Increasing User Acceptance

Within the project Fleets Go Green, a positive user acceptance has been identified as one of the major drivers for the successful integration of electric vehicles into corporate fleets. Perceptions that negatively affect the user acceptance of electric vehicles, such as limited driving distance and higher purchase prices, hinder the wider acceptance of electric vehicle fleets. Thus, overcoming the negative perception of electric vehicles and promoting their positive image is strongly recommended.

Fleet owners and managers should focus on involving their employees in the process of integrating electric vehicles into their fleets. Many people do not have experiences with electric mobility and therefore judge electric vehicles based on rumors. Inviting employees to test electric vehicles and to experience this new driving concept on a first hand basis may be a good starting point for fleet owners. In addition, stressing out the environmental advantages may provide additional motivation for potential drivers to use the vehicles in daily business. Thus, an increased internal communication of the availability of new vehicle concepts may encourage their further acceptance. Furthermore, accommodating simple booking systems (integrated into well-known applications like e.g. Outlook) could increase the adoption of new fleet vehicles in a corporate entity. Easy access to charging facilities and additional parking lots would encourage a positive perception of these fleets, as well. As a result, employees themselves can become promotors for new mobility concepts. Besides fleet owners, vehicle manufacturers can contribute to the user acceptance via addressing possible negative perceptions of fleet owners. To this end, they could address the design requirements of commercial fleet owners, offer specialized equipped vehicles and provide suited solutions to overcome the disadvantage of short driving distances and low battery performances. Furthermore, they could provide extensive consultancy services for prospective buyers. As manufacturers can rely on well-known and documented use cases, they can help fleet owners to choose electric vehicles which fit to the intended use and which do not exceed its capabilities. Political decision makers should support the user acceptance by developing incentives such as bus lines with electric vehicles, additional charging stations or parking lots. Highlighting the environmental advantages of alternative driving concepts may also create a positive influence on potential drivers. Finally, to speed up the growth of electric mobility, political decision makers should also set a good example in becoming first adopters. This would lead to an increased visibility, and an earlier achievement of a critical threshold.

7.3 Defeating Range Anxiety

A major influencing factor on the perception of electric vehicles is their (limited) range. However, the analysis of the use cases within the project Fleets Go Green has shown that short distance trips far outweigh long distance trips. Therefore, an insufficient range for the intended applications was not relevant in the project in nearly all cases. However, range anxiety was still a concrete barrier for some fleet users. Thus, defeating the, in most times, unfounded range anxiety is a major recommendation.

In order to reduce range anxiety, fleet owners should encourage employees to test the electric vehicles and attend trainings. These trainings may focus among others on transferring practical knowledge about how to assess the remaining range of a vehicle and what possibilities the drivers have to extend the range. An empty battery on the way can also be avoided by implementing range-forecasting tools and using them at vehicle booking. Therefore, vehicle manufacturers should provide precise forecasting tools for the remaining range that reflect the diverse influencing factors of the energy demand of vehicles, e.g. weather conditions, number of persons travelling and carried weight. Manufacturers can also benefit from such tools as they can be used for data analysis and differentiating between market segments and/or use cases and providing each segment/case with cost-effective solutions, e.g. with tailored battery capacity. Politics may further encourage that standardized driving cycles provide better results in terms of fuel and energy consumption. This can lead to more confidence in the manufacturers' instructions.

7.4 Optimizing Availability with Intelligent Charging Infrastructure

A viable electric vehicle fleet needs a sufficient charging infrastructure. Aspects like charging times and number of charging stations highly influence the availability of electric vehicles. Therefore, the installation of intelligent charging infrastructure is recommended in order to optimize the availability of an electric vehicle fleet.

Fleet owners should address the charging infrastructure during the planning phase of electric vehicle fleets, as well. This can be done with the help of integrated planning tools that simultaneously address the fleet and charging infrastructure planning. Besides higher charging powers, intelligent charging stations are capable of the controlled charging of vehicles, e.g. charging only at times when electricity from renewable resources can be supplied. However, reducing charging times and thus increasing availability of electric vehicles is not only a concern of fleet owners. Vehicle manufacturers should equip electric vehicles with on-board chargers that are capable of charging on higher charging powers. Furthermore, political decision makers are responsible for creating internationally accepted norms for charging infrastructure for AC and DC charging. While DC charging comes with higher cost, AC charging on higher powers decreases charging times and thus increases the accessibility of electric vehicles.

7.5 Overcoming the Price Aspect

The research results revealed that another obstacle for the increased adoption of electric vehicle fleets is the relatively higher purchase price compared to conventional vehicles. The consideration of the total cost of ownership (TCO) instead of the purchasing price, the offering of customized designs for fleet owners as well as incentives from political decision makers can support the overcoming of this price aspect and support the adoption of electric vehicles fleets.

Fleet owners should not only consider only the purchase price but also the total cost of ownership when the purchasing decision is in question. On one hand, loading the battery with electricity may create a price advantage during the use phase of vehicles over the increasing prices of fossil fuels. In addition, the benefits to the company, due to the positive image impact of acquisitions, should be granted as a long-term interest. On the other hand, residual value risk, for example due to the necessity of additional replacement batteries, may be a negative aspect in the long term. Therefore, the consideration of TCO is a fundamental aspect for electric vehicle fleet managers. Vehicle manufacturer may contribute towards overcoming the price aspect through enlarging their product portfolio targeting the requirements of fleet owners. Design adjustments based on the use case of vehicles, such as scaled down vehicles or avoiding unnecessary additional parts may support the adoption of electric vehicles. In this way, they can achieve an extended range and develop a solution for price obstacles. In addition, they can promote the electric vehicle utilization in fleet operations with additional marketing actions, like providing advice or consultancy services for potential drivers and fleet owners, implementing price reliability or extended warranties. Political decision makers can support the adoption of electric vehicle fleets via charging stations or additional service infrastructure. These additional incentives would pay off when the total societal costs for the public are considered. Political decision makers may as well promote transparency via making the total societal costs quantifiable for potential drivers to influence their decision making process towards the adaptation of electric vehicles and to underline the advantages of electric vehicle fleets. The decision-making criteria for purchasing vehicles are ranked as functionality, cost and ecology respectively. Political decision makers can change this ranking in favor of electric vehicles through some penalties for vehicles with higher fossil fuel consumption or incentives for electric vehicles. Regulations such as speed reduction in autobahn to decrease the fuel consumption may hinder the conventional vehicles utilization or promotions such as enhanced parking facilities may encourage the adoption of electric vehicles.

7.6 Ensuring Environmental Advantages

The main idea of electric vehicles is providing a solution for an environmental friendly mobility. However, electric vehicles are to the current state not always a better choice from an environmental perspective. In order to get a step closer to the vision of electric mobility and ensure the environmental competitiveness of electric vehicles, the common effort of political decision makers, manufacturers and users is needed.

Due to the high number of influencing factors and complex interdependencies, general statements about the advantageousness of electric vehicles in corporate fleets cannot be made. Case based decision-making is required instead and diverse parameters should be considered carefully prior to choosing electric vehicle concepts and matching them to use cases. Fleet owners should choose vehicles in accordance with the requirements of the intended use cases. Among others, geographical features and the regional climate have a strong influence on the energy consumption of electric vehicles. Furthermore, the main environmental impacts in the use phase of electric vehicles stem from the electricity generation. Therefore, electricity from renewable resources should be the desired option when charging electric vehicles. Finally yet importantly, the driver affects the energy consumption of vehicles to a great extent. Therefore, employees should be sensitized to their influencing of the energy consumption of an electric vehicle and be given recommendations how to drive in an energy efficient manner. Low utilization of electric vehicles in fleet applications must be avoided. A higher utilization of a fleet could be achieved through the optimized definition of fleet size and composition. Implementing additional use cases for an existing fleet and thus extending the fleet tasks to a broadened area, e.g. one shared fleet for care service (day use) and facility security service (night use) may be a viable option for fleet owners. Another option to increase the utilization rate is to allow private usage of vehicles outside of working hours. For this purpose, fleet owners may develop new business models that adequately account for the distance travelled and possible differences in the state of charge of battery after returning the vehicle to its base location. Auxiliary power demand (typically heating and cooling) should be kept at a minimum level. A possibility to reduce the energy demand for air conditioning from the perspective of a user is to park vehicles in sun-protected parking lots. In the view of a vehicle manufacturer, measures to cut auxiliary energy demands will become inevitable. To mitigate the effects of ambient temperatures on air conditioning, the interior of the vehicle should be better insulated against heat losses in winter and heating up in summer months. However, in this case further effort for this insulation must also be evaluated and related to the resulting energy reduction. As the battery production dominates the environmental impact of an electric vehicle in the production phase, increasing its resource efficiency is essential to lever the ecological advantages of electric vehicles starting from the early stages of the use phase. To this end, vehicle manufacturers should provide customers with an individually scalable battery capacity for the intended use cases. Modular battery concepts should be developed for this purpose and incorporated into electric vehicles. Electric vehicles may further be integrated into the electricity grid and be used for temporary energy storage. Electric vehicles may be used in this way for supplying factories during peak power demands or storing excess renewable energy during sunny or windy days. Electric vehicles may become this way an essential element for evening out supply and demand in the electricity market. Another interesting scenario to better use the batteries is to extend their life cycle. Vehicle manufacturers could develop business models for using worn out batteries after their first life in an electric vehicle as stationary energy storage for residential photovoltaic systems. A further determining factor influencing the environmental impacts of the battery production, that concerns vehicle manufacturers, is the location of the production facilities. Decision-making prior to choosing a production site should address the potential environmental impacts in production. Moreover, zero impact production should become a relevant factor in the location selection for new production facilities. Furthermore, it is necessary that political decision makers create incentives for the use of electricity from renewable resources and for developing the renewable electricity generation infrastructure. In order to reach the levers of a more effective usage of batteries (as depicted before), political decision makers are under obligation to create a legal framework for the proposed use of electric vehicles as temporary energy storage. Additionally, introducing new regulations, e.g. CO_2 targets for corporate fleet operators, may be an effective way to raise attention for corporate green fleets.